SIMPLICIUS

On Aristotle

On the Heavens 2.10-14

SIMPLICIUS

On Aristotle
On the Heavens 2.10-14

Translated by
Ian Mueller

B L O O M S B U R Y
LONDON • NEW DELHI • NEW YORK • SYDNEY

Bloomsbury Academic
An imprint of Bloomsbury Publishing Plc

50 Bedford Square
London
WC1B 3DP
UK

1385 Broadway
New York
NY 10018
USA

www.bloomsbury.com

Bloomsbury is a registered trade mark of Bloomsbury Publishing Plc

First published in 2005 by Gerald Duckworth & Co. Ltd.
Paperback edition first published 2014

British Library Cataloguing-in-Publication Data
A catalogue record for this book is available from the British Library.

ISBN HB: 978-0-7156-3342-7
PB: 978-1-4725-5812-1
ePDF: 978-1-4725-0115-8

Library of Congress Cataloging-in-Publication Data
A catalog record for this book is available from the Library of Congress.

Acknowledgements
The present translations have been made possible by generous and imaginative funding from the following sources: the National Endowment for the Humanities, Division of Research Programs, an independent federal agency of the USA; the Leverhulme Trust; the British Academy; the Jowett Copyright Trustees; the Royal Society (UK); Centro Internazionale A. Beltrame di Storia dello Spazio e del Tempo (Padua); Mario Mignucci; Liverpool University; the Leventis Foundation; the Arts and Humanities Research Board of the British Academy; the Esmée Fairbairn Charitable Trust; the Henry Brown Trust; Mr and Mrs N. Egon; the Netherlands Organisation for Scientific Research (NWO/GW); Dr Victoria Solomonides, the Cultural Attaché of the Greek Embassy in London. The editor wishes to thank A. Bowen, G. Guldentops, S. Leggatt, and D. Furley for their comments, John Sellars and Inna Kupreeva for preparing the volume for press, and Deborah Blake who has been Duckworth's editor for all volumes since the beginning.

Typeset by Ray Davies

Contents

Preface

The first draft of this translation of Simplicius' commentary on Book 2 of Aristotle's *De Caelo* was finished in 1998-99 when I was an exchange fellow at the Centre d'Histoire des Sciences et des Philosophies Arabes et Médiévales (CHSPAM) at the Centre National de la Recherche Scientifique (CNRS) in Paris.

I would like to thank the personnel, staff, and associates of CHSPAM for their assistance, intellectual and personal, in making my year in Paris profitable and pleasurable, with particular thanks to the then director, Roshdi Rashed and to Tony Lévy, Régis Morelon, Pierre Pellegrin, Muriel Rouabah, and Bernard Vitrac for their interest and kindness. I also thank the University of Chicago and the CNRS for choosing me as an exchange fellow for that year.

Completion of the translation was set back by two years because of administrative duties at the University of Chicago. I am grateful to the readers of the translation for their comments and suggestions, which I have followed more often than not. The names of most of them are not known to me, but I am in a position to mention and especially thank Alan Bowen and David Furley. John Sellars has done admirable editorial work, making many improvements, including the elimination of all too many mistakes. For the all too many that remain I am, of course, fully responsible.

My deepest debt is, as always, to my wife Janel Mueller, who, while pursuing an academic and administrative career of her own, provided me with more support and encouragement than anyone is entitled to.

Details concerning the text translated and some of my translational practices can be found in the Introduction.

Chicago — Ian Mueller

Abbreviations

CAG = *Commentaria in Aristotelem Graeca*, Berlin: G. Reimer, 1882-1909.

DK = Diels, Hermann, and Kranz, Walther (ed. and trans.), *Die Fragmente der Vorsokratiker*, 6th edn, Berlin: Weidmann, 1954.

DSB = Gillispie, Charles Coulston (ed.), *The Dictionary of Scientific Biography*, 16 vols, New York: Charles Scribner's Sons, 1970-80.

El. = *Euclid's Elements*, vols 1-4 in Heiberg, J.L. and Menge, Hermann (ed. and trans.), *Euclidis Opera Omnia*, Leipzig: Teubner, 1883-1916.

FGrH = Jacoby, Felix (ed.), *Die Fragmente der Griechischen Historiker*, Berlin: Weidmann, 1923-.

Guthrie = Guthrie, W.K.C. (ed. and trans.), *Aristotle, On the Heavens*, Cambridge, MA: Harvard University Press, and London: William Heinemann, 1939.

Heiberg = *CAG*, vol. 7.

Karsten = Karsten, Simon (ed.), *Simplicii Commentarius in IV Libros Aristotelis De Caelo*, Utrecht: Kemink and Son, 1865.

Leggatt = Leggatt, Stuart (ed. and trans.), *Aristotle, On the Heavens I & II*, Warminster: Aris and Phillips, 1995.

LSJ = Liddell, George Henry, and Scott, Robert, *A Greek-English Lexicon*, Oxford: Clarendon Press, and New York: Oxford University Press, 1996.

Moraux = Moraux, Paul (ed. and trans.), *Aristote: du Ciel*, texte établi et traduit par Paul Moraux, Paris: Belles Lettres, 1965.

Oeuv. Comp. = Platon, *Oeuvres Complètes*, 27 vols, Paris: Belles Lettres, 1920-64.

PW = *Paulys Realencyclopaedie der Classischen Altertumswissenschaft*, 51 vols, Stuttgart: J.B. Metzler, 1893-1997.

Introduction

1. Simplicius' commentary on Aristotle *On the Heavens* 2.10-14

Richard Sorabji

Aristotle believed that the outermost stars are carried round the earth on a transparent sphere. The sun, moon and planets are carried on different revolving spheres closer to the earth. The spheres are angled in relation to each other, with some counteracting others in such a way as to explain the apparent irregularity of planetary motion in relation to the earth. The spheres and celestial bodies are composed of an everlasting fifth element, which has none of the ordinary contrary properties like heat and cold which could destroy it, but only the facility for uniform rotation. But this creates problems as to how the heavenly bodies produce light and, in the case of the sun, heat.

The value of Simplicius' commentary on the first nine chapters of Aristotle's *On the Heavens* Book 2, already translated by Ian Mueller in an earlier volume, lies partly in their preserving the lost comments of Alexander, the leading champion of Aristotelianism at the end of the second century AD, and partly in Simplicius' objections to Alexander in the sixth century. The two of them discuss not only the problems mentioned, but also whether soul and nature move the spheres as two distinct forces or as one. Alexander appears, in a work preserved in Arabic translation, to have simplified Aristotle's system of 55 spheres down to seven, and some hints may be gleaned as to whether, simplifying further, he thinks there are seven ultimate movers, or only one. Simplicius, by contrast, and his teacher Ammonius, endorsed Ptolemy's hypothesis of a ninth sphere, to account for the precession of the equinoxes (462,12-31).

Sosigenes, the teacher of Alexander, is quoted in Chapter 12 as recording that, because Aristotle's system of spheres was concentric, it could not account for the annual approach and retreat of Venus and Mars in relation to the earth. He alleges that Aristotle recognised this in a lost work. Alternatives to concentric movements were eccentric movements or epicycles. An eccentric movement relative to the earth is a circuit round the earth whose (movable) centre is not the earth's centre. An epicycle is a circle upon a circle. Simplicius thinks that eccentrics may have been introduced by Pythagoreans and Iamblichus ascribes to them the epicyclic hypothesis as well.

Aristotle's system was a development of the earlier systems of Eudoxus and Callippus, and Ian Mueller translates Pseudo-Alexander's treatment of these systems from another work.

Chapter 13 provides a feast for lovers of the Presocratics and other early philosophers. We hear of theories which, unlike Aristotle's, made the earth move, in the case of certain Pythagoreans, around a central fire, in the case of the Platonist Heracleides, by spinning on its axis. For those who saw the earth as stationary, some explained this like Empedocles, who is extensively quoted, by the behaviour of vortices, others by the principle of no sufficient reason for motion. Aristotle had already considered these alternatives to his own explanation in terms of natural places.

2. The text

Ian Mueller

This translation of Simplicius' commentary on Book 2 of Aristotle's *De Caelo* (*On the Heavens*) is based on Heiberg's edition of the text printed as volume 7 of *CAG*. For the text of *De Caelo* itself I have relied on Moraux. Since it seems reasonably clear that Heiberg's edition should not be regarded as definitive[1] and the present textual situation affects my translational practices, I wish here to say a few words about Heiberg's edition. My remarks are based on Heiberg's preface to his edition (cited here by Roman numeral page) and his earlier, more detailed but slightly discrepant report to the Berlin Academy (Heiberg (1892)). I confine my remarks to Book 2, the situation for Book 1 being significantly different.

For Heiberg the most important manuscript is:

A Mutinensis III E 8, thirteenth-fourteenth century, in the Este Library in Modena (Wartelle (1963), no. 1052)

Heiberg (1892), p. 71, singles out A for its correctness and purity. But he admits that it is badly deficient and hastily written, with frequent incorrect divisions of words, misunderstandings of abbreviations, arbitrary use of accents and breathing marks, extremely many omissions, and frequent insertions in a wrong place of words occurring in the vicinity. A glance at the apparatus on almost any page of the edition of Book 2 makes clear how often Heiberg feels forced to depart from A. On the whole these departures seem justified, but, as I shall discuss further below, there are many cases where he follows A and produces a text which seems to me impossible or at least very difficult.

Heiberg thought that A and another text, which he calls B, derived independently from a lost archetype. B stops in Book 1, the remaining pages being torn out. However, there are two other manuscripts which Heiberg took to be copied from B:

J Taurinensis C.I.3, sixteenth century, in the National Library of Turin (Wartelle 1963), no. 2086)

Perusinus cod. 51, fourteenth century, in the Municipal Library of Perugia (Wartelle (1963), no. 1658)

Heiberg makes no use of either of these manuscripts in his edition of Book 2, but it looks as though attention ought to be paid to them.[2]

Four manuscripts which Heiberg does cite are:

C Coislinianus 169, fifteenth century, in the National Library in Paris (Wartelle (1963), no. 1560)
D Coislinianus 166, fourteenth century, in the National Library of Paris (Wartelle (1963), no. 1558)
E Marcianus 491, thirteenth century, in the library of San Marco, Venice (Mioni (1985), pp. 299-300; not in Wartelle (1963))
F Marcianus 228, fifteenth century, in the library of San Marco, Venice (Wartelle (1963), no. 2129)

Heiberg took D and E to be significantly different from A and B, and C to be intermediate between D and E, on the one hand, and A and B, on the other. C and D are, in fact, texts of *De Caelo* with extensive marginalia from Simplicius' commentary (not necessarily word-for-word quotations) rather than complete texts of Simplicius. According to Heiberg E, which is a complete (although lacunose) text, and D were copied from the same prototype, E being copied by an uneducated scribe. E was corrected by Bessarion (E^2), using the Latin translation of William Moerbeke, a work to which I shall return shortly. Heiberg sometimes adopts readings of D or E and, less frequently, of C. Heiberg's treatment of F, which contains only Books 2 to 4, causes the greatest difficulty. Heiberg decided on quite inadequate grounds, that F is a descendant of A. He cites it only where it seems useful, so that, as he says, nothing can be concluded about its contents in places where it is not mentioned in the apparatus.

Heiberg also cites three printed versions of the commentary in his apparatus:

(a) The *editio princeps* of the Greek text: *Simplicii Commentarii in Quatuor Libros de Coelo, cum Textu Ejusdem*, Venice: Aldus Romanus and Andrea Asulani, 1526
(b) The *editio princeps* of the Latin translation of William Moerbeke: *Simplicii philosophi acutissimi, Commentaria in Quatuor Libros De coelo Aristotelis*. Venice: Hieronymus Scotus, 1540
(c) Karsten (1865)

Citations of (a) are rare because Heiberg ((1892), p. 75) realised that it was a translation back into Greek of Moerbeke's Latin translation.[3] However, he did not realise that (b) was 'corrected' in the light of a. Since Moerbeke used an older manuscript of the commentary than any now extant, it is clear that one condition of a satisfactory edition of the Greek text is a new edition of Moerbeke's translation. The situation is made even more complicated by the discovery of a translation by Robert Grosseteste of parts of Simplicius' commentary, including the whole of Book 2.[4] Clearly an edition of this text is needed for the reconstruction of Simplicius' Greek. In my reports on what is in Heiberg's apparatus criticus I usually omit what he says about (b).

Karsten's edition was published one year after his death. It includes no critical apparatus, and has no preface by Karsten. Throughout it is based on single manuscripts, in the case of Book 2 on:

Paris 1910, dated 1471, in the National Library in Paris (Wartelle (1963), no. 1396)

In the absence of a critical apparatus or inspection of Paris 1910, it is impossible to tell what alterations of his source Karsten made, but it seems certain that he made some 'improvements'.[5] In Heiberg's judgement this manuscript is a descendant of:

K Marcianus 221, fifteenth century, in the library of San Marco, Venice (Wartelle (1963), no. 2122)

Books 2-4 of K were copied from F and corrected by Bessarion on the basis of Moerbeke's Latin translation (K^2). Not surprisingly Heiberg makes very little use of K, but his apparatus includes an extensive, although not complete, record of Karsten's text. As I have suggested, the major gap in his apparatus is the incomplete record of F. However, my main subjective impression from reading the apparatus is the large number of places in which Karsten and F agree against the text printed by Heiberg, frequently producing a *lectio facilior* and, in the case of citations or paraphrases of Aristotle, frequently agreeing with what Moraux prints. (In many cases Moraux's apparatus records variants which agree with what Heiberg prints.) I am not always confident about the distinction between a *lectio difficilis* and a *lectio impossibilis*, but I have fairly often chosen to translate a reading from Heiberg's apparatus rather than what he prints. Such divergences are recorded in the Textual Suggestions on pp. 135-6, and in a note at the relevant place in my text. I do not think of most of these changes as emendations, since there is no point in talking about emendations until a better Greek text of the commentary is available to us.

3. Issues of translation
Ian Mueller

In English it is customary to distinguish among sun, moon, the stars, and the planets. The usual term for referring to all of them together is 'heavenly body'. The usual Greek word for all of them is *astêr* or *astron*. I have chosen to translate these two words as 'star' rather than 'heavenly body' because Simplicius often uses that phrase (*ouranion sôma*) and frequently uses it to refer not to what we would think of as a heavenly body, but to the body of the whole heaven or a heavenly sphere. So the reader has to bear in mind that in this translation the general meaning of 'star' is 'sun, moon, star, or planet'.

In Greek astronomy the word 'planet' (*planêtês*, usually plural) refers to the moon, sun, and five known 'planets', Mercury, Venus, Mars, Jupiter, Saturn. 'Planet' occurs relatively rarely in the commentary, usually in connection with other writers or when Simplicius is quoting a passage in which Aristotle uses the word; see the Greek-English Index under *planêtês*. In the translation I have used the customary names for the five planets, although, of course, the Greeks associate them with their own divinities, Hermes, Aphrodite, Ares, Zeus, and Kronos. In the commentary on Chapters 10 to 14 Simplicius explicitly refers twice to Venus as the star (or so-called star) of Aphrodite (496,6 and 504,27) and twice to Mercury as the star of Hermes (474,20 and 495,26); Simplicius also refers to Venus as the star of *Heôsphoros* at 495,26, where I have given the translation 'Morning Star'. In addition, there are several implicit references of this kind, including two in Aristotle at 292a5 (Mars), which is quoted by Simplicius at 479,16 and paraphrased at 481,10, and *Metaphysics* 12.8, 1073b34-5 (Jupiter and Saturn), which is quoted by Simplicius at 497,11. Other examples of such implicit references are found at 471,17 (Saturn), 474,21 (Venus), 495,27-8 (Mars, Jupiter, Saturn), 496,7-8 (Mercury, Mars, Jupiter, Saturn), 497,23 (Venus, Mercury, Mars), 498,11 (Jupiter), 498,18 and 25 (both Saturn), 504,28-9 (Venus, Mars). But Simplicius also refers to the planets directly with the name of the appropriate divinity; see 474,17-19 (Venus, Mercury, Jupiter), 474,24 (Venus), 491,21 (Jupiter, Saturn), and 502,28-30 (Venus, Mercury, Jupiter, Saturn). And he has no hesitation in speaking about the sphere or spheres or the distance of, say, Venus (references in the Index of Names). So it looks as though Simplicius is willing to refer to the five planets simply with the name of the corresponding divinity. Sun and moon are always referred to directly as *hêlios* and *selênê*. I do not capitalise 'sun' and 'moon'; neither does Heiberg, except sometimes in contexts where Simplicius mentions other planets as well (471,6-18, 474,18-25, and 495,28 (where Simplicius mentions that Saturn was once called 'star of *Hêlios*')).

The word 'planet' comes from the verb *planasthai*, meaning 'to wander'. Simplicius uses *planasthai* either as an infinitive or as a present participle. As an infinitive it always occurs in an expression like 'star (or sphere) which is said (or thought) to wander', an indication of Simplicius' Platonist belief that the planets don't 'really' wander. When it is used as a participial adjective and the noun it modifies is either explicit or it is reasonably clear what the noun is I have translated the participle 'planetary' except when the noun is 'star', in which case I have translated 'planetary star' as 'planet'. When the participle is a substantive there are more difficulties. In these cases I have rendered the plural as 'planets'. However, Simplicius often uses the singular to refer to the domain of the planets; here I have rarely (only at 435,3) used 'planetary' and chosen 'the planets' instead.

I have translated the word *aplanês* (unwandering) as 'fixed'. Simplicius most often uses this adjective by itself without a noun. Most frequently it is in the feminine singular and the noun to be supplied is 'sphere' (*sphaira*); so 'fixed sphere' in this translation refers to what we ordinarily call the sphere of the fixed stars. Sometimes 'fixed' is in the plural and the noun to be supplied is 'star'. I list here all the places where *aplanês* modifies a noun:

fixed heavenly body (= the fixed sphere): 408,3
fixed heaven (= the fixed sphere): 420,36; 444,28; 459,26; 487,14
fixed stars: 415,22; 444,18.28; 453,12; 454,25; 455,9; 490,27; 537,14; 549,5 (fixed star at 445,15)
fixed sphere: 449,2; 453,14; 514,16; 462,13; 548,27
At 455,5 the words *aplanê tôn astrôn* (the fixed among the stars) is quoted from Plato

Aristotle does not use the word *aplanês* in *De Caelo*. He normally uses *endedemenos*, which I have translated 'fastened'.

The word *to meson* is sometimes translated 'middle', but it has frequently seemed to me necessary to translate it 'centre'. Because of this fact I have chosen to translate the word *kentron*, usually translated 'centre' as 'central point'.

A considerable portion of the commentary is very close paraphrase of the text being commented on, so close that the line between paraphrase and quotation is frequently difficult to draw. Heiberg indicates what he takes to be quotations by spacing letters more widely; I use single quotation marks. In general I have identified fewer quotations than Heiberg does, and have tried to restrict quotation marks to strings of at least five or six words which are strictly identical with the corresponding Aristotelian passage. The choice is to some extent arbitrary since Simplicius sometimes omits particles which are grammatically necessary in an Aristotelian sentence but unnecessary when the sentence is embedded in a larger context, and he sometimes merely varies Aristotle's word order or inserts a word which is only implicit in the text he is dealing with.

Notes

1. See especially the unpublished thesis, Hoffmann (1981), e.g., p. 18. I have relied heavily on this work.

2. So Hoffmann (1981), pp. 240, 293.

3. A fact first noticed by Peyron (1810).

4. See Allan (1950), and Bossier (1987), pp. 289-90 and 320-5.

5. See Bergk (1883), p. 143, n. 1 and p. 148.

Simplicius

On Aristotle

On the Heavens 2.10-14

Translation

On the Second Book of Aristotle's On the Heavens

[Chapter 10]

291a29-b10 Let their <the stars'> order – [the way in which each moves, with some being prior others posterior, and the way in which they are related to one another in their distances – be studied on the basis of astronomical works, since they are spoken about sufficiently <there>.

291a32 And it turns out that the motions of each are in proportion with its distance, some being faster others slower; for since it is assumed that the last revolution of the heaven is simple and the fastest and the revolutions of the others are slower and more multiple (because each moves on its own circle in a direction contrary to the heaven), it is consequently reasonable that the one which is closest to the simple and first revolution traverse its circle in the most time, the furthest away in the least time, and in the case of the others, the closer always in more time, the further away in less. For the closest one is most dominated <by the last revolution>, the furthest away least of all because of the distance. And the intermediate ones <are dominated> in proportion to their distance,] as the mathematicians also prove.

It was necessary for those who discuss heavenly things also to discuss
the order with respect to position of the spheres and the stars and to 470,30
say which ones are prior and more proximate to the fixed sphere, and which are posterior and closer to earth, and further how they relate to one another in their distances, taken with respect to the earth, on
the basis of which the ratios of their sizes are apprehended. Let these 471,1
things, he says, 'be studied on the basis of astronomical works'. For there, demonstrations have been given about the order of the planets and their sizes and distances. Eudemus recounts that Anaximander was the first to have given an account of their sizes and distances, and
he credits the Pythagoreans with the first ordering of their position.[1] 5
The sizes and distances of the sun and moon have been figured out before now, the first impulse to their apprehension being taken from eclipses (and it is likely that Anaximander also discovered these things). And the first impulse for the apprehension of the sizes and distances of Mercury and Venus was taken from their conjunction[2] with the sun and moon. These sizes and distances were made more

precise by people after Aristotle, and most completely by the associates of Hipparchus, Aristarchus, and Ptolemy.[3]

(291a32) He says that it turns out that the motions are in proportionality with their distances because those close to earth, like the moon, move faster, those further from the earth slower in proportionality with their distances. It was appropriate to have added this point to the discussion of order and distances, but it rightly raised a difficulty: why the <stars> near the earth move faster, and the ones which are higher and nearer to the fixed sphere more slowly; for example Saturn is restored in thirty years, whereas the moon makes a revolution in a month.[4] And the difficulty might be stirred up from two sources. First, from size, since a greater body moves with its own motion faster, as Aristotle himself has said,[5] and what contains is always greater than what it contains. How, then, is it the case that the outer motions are not faster in proportion with their size and distance, but on the contrary slower? And, second, it is also necessary to raise the difficulty on the basis of proximity to or distance from the fixed sphere, since if the fixed sphere has the fastest motion of all the spheres, it follows that things nearer to it move faster than things further away in proportion with their distance, and if the earth is naturally motionless it should have been the case that things nearer to the earth be slower than things further away, and this again in proportion with their distance.

He solves these difficulties skilfully by saying that, since the fixed sphere has a single motion from the east, which is the fastest, and the planetary spheres have this motion and an opposite[6] one, it would be reasonable that the closest to the fastest revolution traverse a revolution opposite to it in the most time, because it is dominated and resisted by the fastest revolution, whereas the furthest away is dominated least of all because of the distance and moves faster than the others, 'and the intermediate ones <are dominated> in proportion to their distance, as the mathematicians also prove'.

What then? Do the spheres nearer to the fixed sphere move more slowly because they are forced to by that sphere? However, if they are moved by force, they also move in a completely unnatural way. Consequently both of their motions, that from the east in which they are carried around with the fixed sphere and their own motion from the west would be forced and unnatural.[7] Alexander[8] confronts this difficulty in a good way. He says:

> The fact that the motion of the fixed sphere is fastest is a cause of the sphere of Saturn having a slower restoration, but the sphere is not unwilling. For these things would be chosen and willed, since nothing would be better for them or more choiceworthy than this cosmic order. So, the cause which works in terms of necessity and the cause which works in terms of the

best have come together, since it is necessary that there not be force only. And because it is best that things be this way Saturn moves voluntarily; but because it is near to the sphere moving in a contrary direction, it does so by necessity. For these interactive motions are not unnatural for the spheres since they have no unnatural motion because their motion doesn't even have an opposite.[9] But all their motions will be natural for them in that some stem from themselves and some from their interaction. Consequently one should also say the same thing in the case of the motion which they have because of being moved along with the fixed sphere: it, too, is not unnatural for them.

But perhaps the difficulty still remains. For let it be the case that their interactive motions are neither forced nor unnatural but voluntary. Would it still really be necessary that the spheres, which have soul and share in action (as he will say they do),[10] have their own motions in a completely natural way? If they have two motions, one from the east and one from the west, since their motion from the east belongs to the fixed sphere (for they have the motion because they are carried with the fixed sphere), and they have the motion from the west and it is dominated and resisted by the fixed sphere, what proper motion can they have naturally? So Aristotle's account does not solve the difficulty of how it can still be true that a greater body moves with its proper motion faster and how what is adjacent to the fixed sphere – which has the fastest motion – and is obviously more akin to it (since it has been given the neighbouring place in accordance with its kinship in substance) has a slower motion whereas what is proximate to the earth, which is motionless, has a faster one. So I do not think <Alexander> has solved these difficulties; rather he has thought of another cause which does not get away from force completely. And even if they have this derivative motion from the east by being moved along with the fixed sphere, nothing prevents their moving this way voluntarily because they also have their own motion in accordance with their own unhindered impulse and proceed naturally as if they were not carried around with the fixed sphere.

But if their own natural motion is resisted and dominated, how can that not be forced? Unless someone were to say the following:[11]

<Spheres> which are near to the fixed sphere, insofar as they are akin to it, have both their own motion and their motion from the east; and the greater sphere always moves faster <from the east>, the size and the speed of the motion being in the same ratio, there being one nexus of all the spheres in a single heavenly body; but insofar as they have a nature which makes them move in a contrary way, the spheres which move a little way under <the fixed sphere> do move faster with the motion

which is akin to the fixed sphere because they remain more in the special character of that sphere, but they move with the motion of the contrary nature more slowly[12] because in a way they do not stand still completely in relation to that sphere, just as the sphere of the moon (which is further away from the fixed sphere, not just spatially but also in substance, and closer to what comes to be) since it is smaller, moves more slowly with the motion of the fixed sphere, but faster with the counter revolution. It is as if you were to conceive some structure of substance in the sublunary world changing from air to water: for the part of the structure which moves a little way out of the air moves with the motion akin to air (the motion upward) faster than the parts of the air which move out further, but it moves downward more slowly, and in succession they have their speed and slowness proportional to their kinship with the air, force not being shown anywhere, but nature itself having each thing. But here <in the sublunary world> this sort of mixture of substance has its existence in terms of change and opposition, there it has it in terms of procession and declension and change of form involving no opposition. For it has been proved that the motion from the east and the motion from the west are not opposites, so that – if this account has anything true to say in connection with these most difficult matters – it is presumably also possible for the same thing to move in both these ways at the same time in accordance with a single nature which is constituted on the basis of procession; for in this way the proportionality of the size to the speed from the upper to the lower will be preserved as in a single whole, and again the motion of the planetary <spheres> as planetary, which itself is also proper <to them>, will no longer have a proportionality of speed in accordance with size, but in accordance with revealing to a greater or less degree the special character of the planetary.

Alexander also confirms that the greater spheres are faster in accordance with their own nature, but that the higher spheres move more slowly because hindered by the fixed sphere from the fact that, as he says, the spheres of Mars and Mercury, which are higher, as he says, than that of Venus and therefore also greater are restored at the same speed as each other and as the sphere of Venus; for the smaller spheres, which are not equally hindered by the last revolution because of the distance, move at the same speed as spheres greater than they are.[13]

But to say that the sphere of Mercury is above that of Venus is either a scribal error in which Mercury is substituted for the sun or it is expressed in accordance with the opinion of earlier people. In the *Republic* Plato describes the celestial sphere in accordance with this opinion;[14] he says that the circle of Venus is sixth <counting> from above and second in brightness after Jupiter, the sun is seventh and the moon eighth, so that he orders Mercury above Venus. Observations in which Mercury is recorded moving under Venus make clear that Mercury has been apprehended under Venus.[15] And this has also been proved by the account of the distance of their apogees and perigees; for the greatest distance of Venus has been proved to be more or less the same as the distance of the sun, so that Venus is nearer than the sun, but the greatest distance of Mercury has been proved to be quite close to the least distance of Venus, and the greatest of the moon close to the least of Mercury – these things are proved in the *Syntaxis* of Ptolemy, if the account of the eccentricity of the stars is transformed into an account of their <distance>[16] from the central point of the earth. But since, as I have said, this is either a scribal error or expressed in terms of an earlier way of describing the sphere, it does not require much discussion.

Alexander also gives another explanation for why the things which are nearer to the fixed sphere are restored more slowly, namely that the higher spheres are greater.[17] And it is clear that containing spheres are greater than contained ones, but, unless the ratios of the distance and the sizes there are apprehended, it is not possible to say 475,1
that the speeds are proportional to the sizes. For the sphere of Saturn is restored in thirty years, that is, in three hundred and sixty months; and let it be assumed that the moon is restored in roughly one month; so, if the size of the sphere of Saturn were more than three hundred and sixty times that of the lunar sphere, it would be possible to declare that the sphere of Saturn moves faster than the lunar sphere; for it is necessary that what moves a greater distance in an equal time move faster – especially in the case of things which move uniformly.

Not only Aristotle but also Plato thinks that what moves in smaller circles moves faster than what moves in greater ones. For he says in the *Timaeus*:

> <The motion of the planets> is along the motion of the Different,[18] which is oblique and goes through the motion of the Same[19] and is dominated by it.[20] Some move in greater circles, some in smaller, and those moving in smaller circles revolve[21] faster, those in greater slower.[22]

And in the *Republic*, when he speaks about their order and makes the fixed sphere first,[23] the moon eighth, and puts the others in between, he adds,

> Of these spheres the eighth was[24] fastest, second[25] and together with one another were the seventh, sixth, and fifth, third moved …[26] the fourth, fourth the third, and fifth the second.[27]

But Plato could be saying that the lower ones move faster because he is not considering the ratio of the size but only the time of restoration, because they are restored in less time.[28] For if, as has been said, the ratio of the size is greater than the ratio of the time of motion, what is restored in less time is slower. Indeed, it seems that Aristotle finds the solution of the difficulty according to which the motion itself[29] of things closer to earth is faster in this way. For if being dominated and resisted by the fixed sphere hinders the motion itself and makes it slower,[30] it is clear that in itself and not in terms of its restoration, the motion of things closer to earth is slower – unless one should say on this account that, although a greater revolution is now faster and able (insofar as it is just up to itself) to be restored together with a smaller one, the domination of the fixed sphere does not make it appear so much faster. And Aristotle provides the reason for the fact that it is not without qualification true that the spheres near to the fixed sphere are slower, but that they appear to be slower than they are: to the extent it was up to themselves they might perhaps be restored together with smaller spheres, but they fall behind being restored simultaneously to the extent of the domination of the fixed sphere; for in this way the doctrine that greater things move faster when they move naturally and as much faster as they are greater remains unshaken. And there is nothing anomalous about there being in the case of some form a suitability of such a kind that it is this thing in itself but it becomes such and such else because of the domination of something stronger, just as it[31] has a limited power in itself but it is and moves *ad infinitum* because of the unmoving cause.

Those[32] who make the assumption that all the spheres have the same motion, that from the east, in such a way that in a day the sphere of Saturn is restored simultaneously with the fixed sphere with a small deficiency, the sphere of Jupiter with a larger one, and so on, escape many other difficulties; for <then> the motion will have speeds proportional to magnitudes and things which are of the same substance will have the same motion.[33] But a hypothesis of this sort has been proved to be impossible. For it is necessary that the revolution of the planets be circular and always the same so that their motion will be ordered and therefore also apprehensible. And will they say that this circle on which they maintain that each of the planets makes its motion from east to west is one of the circles parallel <to the equator> or an oblique circle? For if it were one of the parallel circles the planets could not move further south or north, nor could they rise and set at different places on the horizon. But if it were oblique, it would be necessary that each day each of the planets appear to move southward or northward[34] because, as they say, they all make a revolution on their oblique circle in accordance with each revolution of the universe (except for the degrees by which they are observed to be left behind). But both of these <possibilities> conflict with clear <facts>.

It is worth noting that a difficulty concerning the isodromic stars[35] still remains on every hypothesis: why are the spheres which contain and are contained (which is the same as to say, the greater and smaller spheres) restored in an equal time. For whether one makes the assumption that the fixed sphere and the spheres of the planets move in the same direction or that those spheres which are near to the fixed sphere and are dominated by it move more slowly, one does not preserve the proportion of magnitudes to speeds in the case of the isodromic spheres whether <one says that> those nearer to the fixed sphere move faster or that the smaller ones do.

[Chapter 11]

> **291b11-17** One would most reasonably assume that the shape of each of the stars [is spherical. For since it has been proved that they are not of such a nature as to move on their own, but nature does nothing unreasonably or pointlessly, it is clear that it has given to these unmoving things a shape which is least kinetic; but the sphere is least kinetic because it has no organ for motion; so] it is clear that they will be spherical in bulk.

He has already said that the stars are spherical because they are of the same substance as the heavenly body, and proved that they do not change place because they are spherical, or rather he assumed that they are spherical as a hypothesis, and therefore said, 'furthermore, since the stars are spherical, just as others also say in agreement with us'; however not simply as a hypothesis, but he reasonably used the causal conjunction 'since', and he added the easier justification with the words 'since they generate them from that body'.[36] There he called to mind their sphericity by referring to motion; here he proves in a primary way that the stars are spherical, using two arguments of which the second is double.[37] The first is based on the fact that the stars do not move on their own. He means by this change of place on their own; and forward motion is of this kind. He again[38] takes as an axiom that nature does nothing unreasonably and, having demonstrated already that the stars do not change place on their own, he produces a syllogism which amounts to this:

> (i) The stars do not change place on their own;
> (ii) things of this kind have no organ for this kind of motion because nature does nothing unreasonably;
> (iii) things which have no organ for motion are spherical because they have no extrusion;

(iv) so it is clear that the stars will be spherical in bulk, that is, in body.

But if he proved previously that the stars do not change place because they are spherical by considering on the basis of a division the motion which is appropriate to spheres,[39] and now he proves that they are spherical on the basis of the fact that they do not move, how can the proof not be circular? In response to this they[40] say that he did not prove their lack of motion only on the basis of their being spherical nor their being spherical only on the basis of their not moving; rather both have been proved by several arguments. And for this reason, Alexander says, the proof is not circular.

But how does the fact that the same conclusion has been established by means of other arguments make *this* demonstration not circular? The fact that the statements have been demonstrated through other arguments and not just in this circular way is evidence that and an explanation why the statements cannot be refuted, but how can it be evidence that or an explanation why these proofs are not circular? So perhaps Aristotle is taking sphericity and not having an organ for change of place (from which not changing place follows necessarily) as convertible and demonstrates each from the other in a reasonable way. It is as if someone inferred having milk from having given birth and having given birth from having milk, or inferred human being from mortal, rational animal and inferred the definition from human being. Circular proofs of this kind should not be rejected.[41]

One should also recognise from these things what kind of motion Aristotle denies the stars to have, namely the one which is foreign to spherical shapes, change of place by means of organs. For he says that spherical shape is least kinetic in this way, and he adds the reason, 'because it has no organ for motion'. But he has said that internal motion is most proper to spherical shapes, not just to the heaven, but also to the stars, when he wrote, 'therefore it would seem reasonable that the whole heaven and each of the stars be spherical. For the sphere is the most serviceable shape for internal motion (for in this way they can move most quickly and most perfectly occupy the same place), but it is the least serviceable for forward motion; for it is the least similar to things which move on their own since none of it is detached or projecting, as in a rectilinear figure ....'[42] And what he says in our present text harmonises with this statement since Aristotle says in the present text that the stars do not make their apparent change of place on their own and in the statement he clearly assigns them an internal motion as being appropriate to a spherical shape. And so he says both of these things about spherical shape: that it is least kinetic for changing place on one's own; that 'the sphere is the most serviceable shape for internal motion'.[43]

291b17-23 Furthermore,[44] one and all are similar; [but the moon is proved to be spherical by observation, since, were it otherwise, in its waxing and waning it would not usually be crescent-shaped or doubly convex and only once bisected.

291b21 And again it is shown in astronomy that eclipses of the sun would not be crescent-shaped <if the moon were not spherical>. Consequently, since one star is spherical, it is clear] that the others will be as well.

This is the second argument proving the sphericity of the stars; and it uses an axiom which says that any one of the stars and all of them are similar in shape, since all are of the same simple substance. So, if the moon is proved to be spherical because of its visible illuminations, it is clear that the other stars will also be spherical. For, if it weren't spherical but, say, drum- or lentil-shaped, he says that its illumination in its waxing and waning would not be such as to be usually crescent-shaped or doubly convex and only once bisected.

But if he is calling the full moon bisected, in the way in which Aratus calls it month-bisecting[45] because it divides the month in half, then everything else would be in harmony as would its frequently appearing to be crescent-shaped (since it does so in waxing and waning) and similarly for appearing to be doubly convex. But since shortly hereafter he calls the moon bisected in the way we usually use the word when he says 'for we have seen the moon, when bisected, move under Mars, which was hidden from sight by the dark part of the moon and then moved out from the light and bright part',[46] they[47] correctly interpret the words 'once bisected' as follows: in its waxing and waning <the moon> is crescent-shaped and doubly convex most of the time (for more and less are applicable to these shapes), and, even if it is also bisected both when it waxes and when it wanes, nevertheless this does not happen during a period of time (nor are more and less applicable to the shape), but the time of it is instantaneous, as the word 'once' indicates.

These shapes of the illuminations are, in fact, peculiar properties of what is spherical, because a hemisphere is always illuminated; so when the moon is moving under the sun and is at the same degree <of longitude> as it, the part of the moon which is towards the sun is illuminated and the part facing us is covered in shadow, but when the moon is at a distance from the sun what is illuminated is always a hemisphere, and as much as the hemisphere facing away from us is always losing <illumination>, so much is the part facing us taking it on. Therefore, the moon appears to be crescent-shaped until it is a half-moon, but when half of the upper[48] <lunar hemisphere> and half of the <hemisphere> which faces us are illuminated, the moon is seen as bisected; but from the time when it stands at a square distance from the sun[49] until it is in opposition to it, it is seen as doubly convex.

But when it is in opposition the whole hemisphere which is towards us receives illumination, but what faces up is not illuminated; and again, as it approaches the sun, it first presents itself to us as doubly convex, then as bisected, then as crescent-shaped and then, when it is in conjunction <with the sun>, as covered in shadow. The reason is, as I have said, that a hemisphere of the moon is always illuminated because it is spherical.[50]

Notice that if the moon were drum- or lentil-shaped, it would appear the same as it does now in its conjunctions <with the sun> and at full moons, but if it were at any distance from the sun in either direction, it would not be crescent-shaped or bisected or doubly convex; rather, if it were drum-shaped, the part towards us would be illuminated as a whole because none of it would resist the rays <of the sun>; but if it were lentil-shaped, and its height at the middle were slight, the shape of the illumination would be different.[51]

(291b21) Next he adds another demonstration based on astronomy, namely 'that eclipses of the sun would not be crescent-shaped' as they are now seen to be, if the moon moving under the sun were not spherical. For it has been proved that, when one sphere is darkened by another, the sections are of this kind. But perhaps the other roundish figures such as the drum- or lentil-shaped make crescent-shaped sections when they cover <the sun>. But if they are assumed to move around their own central points, drum- or lentil-shaped things will not make the sections to be of this kind[52] in every position.

[Chapter 12]

291b24-292a18 There are two difficulties[53] [which anyone might reasonably discuss; and we should try to state what appears to be true, considering such zeal to be a form of modesty rather than of rashness, if a person because of his thirst for philosophical knowledge is content with a little understanding in the case of things about which we have the greatest difficulty.

There are many such difficulties and not the least amazing of them is this: what is the reason why, although those things which are more distant from the first motion always have several motions, the things in between have most? For it would seem to be reasonable that, since the first body has one motion, what is nearest to it would have the fewest motions, perhaps two, the next thing three, or that there be some other such order. But, in fact, the opposite turns out to be the case. For the sun and moon have fewer motions than some of the planets but the planets are further away from the centre <of the cosmos> and nearer to the first body. And in some cases this has even been made clear visually; for we have seen the moon, when bisected, move under Mars,[54] which was hidden from sight by the dark part of the moon and then moved out from the light and bright part. The early Egyptians and Babylonians say the same thing about other stars, and they have made observations for very many years, and we have much trustworthy evidence about each of the stars from them.

(292a10) So one might rightly raise this difficulty and also the following: what is the reason why the multitude of stars in the first motion is so great that their whole grouping is thought to be uncountable, whereas for each of the others, there is one separate <star>, and one does not observe two or more fastened in the same motion?

(292a14) It[55] is good to seek even more understanding of these things, even though we have little to start from and are at such a great distance from what occurs in their case. Nevertheless, for those who study on the basis of things of the following kind,] the present difficulties will not seem to be anything inexplicable.[56]

He proposes two remaining difficulties about heavenly things; and they really are the most difficult. The first of them is this: why, when the fixed sphere has one motion, does what is nearest to it, that is, the sphere of Saturn, not have the fewest motions, perhaps two, what comes after it three, or there exist some other analogous numerical order so that what is further away always has more motions, but rather the opposite has turned out to be the case?[57] For the sun and 481,1
the moon are lower than the other <stars> (since he, like Plato,[58] hypothesises that the sun is directly above the moon), but they 'have fewer motions than some of the planets'. For among the planets the motion of the sun is simplest, and that of the moon is simpler than the others, but it should have been the case that higher things which are further away from the centre and nearer to the fixed sphere (which he calls the 'first body') have simpler motions than the sun and the moon. And he shows that the moon is lower than the other <planets> by reference to its recorded passages under <other planets>, one of which he says he has seen, its passage under Mars. He says that he has seen the moon, when bisected, move under Mars, which was hidden from sight by the dark part of the moon and then moved out from the bright part (since the moon was bisected in the course of waxing). But he observed this himself, and Egyptians and Babylonians observed the same thing happening in the case of other higher stars, as has been conveyed by many of their observations concerning each star.

(292a10) Next he adds the second difficulty: what is the reason why the multitude of stars[59] in the fixed sphere is so great that it is thought to be uncountable, whereas one does not observe more than one star in any of the spheres under it?[60]

(292a14) And then, seeing the hazardousness of the inquiry and reckoning it to be frightening because of the magnitude of the difficulties, he makes an exhortation by saying that it will be good to investigate these things and to attain greater understanding or, rather, to demand greater understanding.[61] Alexander thinks that the text is somewhat deficient in this way because he thinks that what is added to it is smoother. But perhaps the text is not deficient. For, even if Aristotle is terse, he does not usually express himself elliptically. Rather what he says is those who have most understanding – and not people taken at random – should investigate such things and not be fearful, even if they have little concerning them to start from and they stand further from what occurs in their case than is their spatial distance from them, as he says elsewhere.[62] Nevertheless, even if this is the way things are, on the basis of the reasoning which will be given, the present difficulties would not seem to be inexplicable.

482,1 **292a18-b10** We think about these things as if they were just bodies,[63] [monads in a certain order and entirely without soul. But we ought to conceive them as sharing in action and life; in this way what occurs will not be thought at all paradoxical. For it seems that (i) the good accrues without action to what is in the best condition, (ii) with a single slight action to what is closest to the best condition, and (iii) with more for what is further away, just as in the case of the body, one is in a good condition without exercise, another by walking a little, a third needs running, wrestling, and getting down in the dirt,[64] and again this good will not accrue to a fourth person, no matter how much he works, although a different good will.[65]

(292a28) It is difficult to be successful either in many things or often; for example, to throw ten thousand Chians[66] at dice is unbelievable, but to succeed in one or two cases is easier. And again, when it is necessary to do one thing for the sake of a second, that for the sake of a third, and so on, it is easy to succeed in one or two steps, but to the extent that more are required it is more difficult. Therefore, one should suppose that the action of the stars to be like that of animals and plants. For here humans perform the most actions, since it is possible to attain many good things, so that humans perform many actions and some for the sake of others. (However, no action is necessary for what is in the best condition,[67] since it is that for the sake of which <everything else is done>, since[68] action is[69] of two kinds, when it is that for the sake of which and when it is for the sake of something.) And indeed other[70] animals perform fewer actions, and <the action> of plants is perhaps one little thing, since either there is some one thing which they may achieve (as is also the case with a human)][71] or the many things they do are all a means to achieving the best thing.

What has been said up to now has concerned the two difficulties, and now, starting after the solution of the first difficulty, he first states the reason why the issue seems so very difficult and says that it is not because of what is sought but because of the people seeking. For we suppose the difficulty to be unsolvable because we think about heavenly things as if they were just bodies without soul and like the monads in numbers, which just have an order relative to one another but are 'entirely without soul'. For the difficulty would be unsolvable if they were this way and no starting point for a solution were discovered by considering them. But we should think about them as having soul, a rational soul, so that they also share in action and a life of action. For we also speak about doing or making in the case of irrational souls and bodies without soul, but we predicate action uniquely in the case of rational souls. If we do conceive them in this way what occurs in the case of the motion of the heavenly bodies will not be thought at all paradoxical. For since these things engage in action, and all action results from motion for the sake of the good, it is clear that those things which possess the best or are the good itself or are united in substance with the good itself (as the prime mover is – and such is the much-honoured intellect) are without action and motion and they possess the good. Or, as he says, one thing has it and another shares in it directly, and the good accrues with a single slight motion to what is closest <to the good>, as in the case of the fixed sphere, and with more <motions> for what is further away, as in the case of the planets,[72] and other things cannot attain it immediately but are contented with being near to things that do attain it, as is the case with the earth, which is also motionless as a result, or[73] also everything sublunary, since the motion of some of them is in a straight line, because they are incomplete, and that of others, fire and the upper air, is circular along with the heaven.

Next he uses the body and health as an example. He says that one body is in good condition independently of exercise because it is in the best condition (this is analogous to what is motionless),[74] 'another by walking a little' (which he compares to the fixed sphere), but a third needs more exercise for being healthy, e.g., running and wrestling and gymnastic workout in wrestling (getting down in the dirt is this 483,1
kind of thing, because wrestling is practised in dirt) – this has been taken as analogous to the planets; but the unmixed goodness of health does not accrue to the body which is least well disposed no matter how much it works (he compares this to the inability of the sublunary to participate immediately in divine goodness and its consequent lack of motion on its own).

(292a28) He perceives that what he has said is still deficient, since he has not stated the reason for the differentiation of the planets – why the sun and moon have fewer motions, the higher planets more; and so he fills out what is left out when he says that more honourable beings perform more actions because it is possible to attain many good things, and that being successful either in many things or often, which is extremely difficult, is more suitable for these beings, 'for example, to throw ten thousand Chians' in playing dice – or Coans, since the text is also written this way because there are large dice on both islands;[75] but it is not just difficult, it is also impossible to do this, 'but to succeed in one or two cases is easier'. And attaining the most complete good through more <steps> is suitable to stronger things, for example, if it is necessary to do one thing for the sake of a second, the second for a third, and the third for a fourth – as it is necessary to learn reading and writing in order to be able to participate in mathematics and to do this in order to do philosophy, and to do this for assimilation to the divine.[76] For even for a weaker person 'it is easy to succeed in one or two steps, but to the extent that more are required it is more difficult'. So, just as among mortal animals the actions of humans are most numerous because it is possible for a human who performs many actions, acting politically for the sake of others and taking their good upon himself, to attain many good things, one should also believe that the action and motion of the stars is many times more variegated, when one is compared with others, because they are able to attain more good things; for a human being is also more honourable than the other animals because it can perform more actions.

The whole argument would amount to the following. If things that move in more ways are more honourable, they move in more ways because there are more ways to be successful; but if they are worse, they move in more ways because they are not able to attain the best by means of a simple motion. Consequently, even if opposites belong to the same things and the same things to opposites, we will not be at a loss for a solution, but we will provide the explanations in a way which is appropriate to the subjects. So, when Aristotle says these things, he is not judging the worth of the gods, since saying such things is precarious, but he is giving starting points for a solution; 484,1 following up on them, we will not be surprised if more honourable things are less active or if worse things are.[77]

Having said that humans perform many actions and in such a way as to perform one action for the sake of another, he adds, in order that someone not take this behaviour as the best, 'however, no action is necessary for what is presumably in the best condition':[78] and he adds the reason, or rather the entire demonstration when he says that what is in the best condition is that of which the essence (*einai*) is to be that for the sake of which, since the best is the end of everything and everything is for the sake of it, but what performs an action is different from that for the sake of which it performs the action. And again he adds the reason for this premiss when he says, 'since action is of two kinds, when it is that for the sake of which and when it is for the sake of something'. For if everything which performs an action performs the action it performs because of a desire for a good, the good will be one thing, what acts another. It follows in the second figure that what is in the best condition has no need of action, since what is in the best condition is that for the sake of which, and what performs an action is not that for the sake of which.

And having spoken in between about the best he connects the words 'and indeed other animals perform fewer actions' and so on with what was said before about humans. He says that the action of plants is one little thing – perhaps <he means> action involving nourishment[79] – since they cannot be successful in many things. He uses the word 'action' in a more ordinary way to apply to the activity of a plant, since action in the strict sense is rational activity.

The next words 'since either there is some one thing which they may achieve' are perhaps not spoken about plants but spoken universally about everything which performs actions[80] because either there is some one thing open to what performs actions which it may achieve (just as in the case of a human being there are more things open to it); or, if what is open is not one thing but several as is also the case with a human, then, nevertheless, these many things are 'a means to achieving the best thing' because everything else inclines towards that and is chosen because of it.

The words 'since either there is some one thing' may also apply to plants with respect to what is expressed by 'perhaps one little thing'; for either there is one good for a plant which it may achieve, just as a human being may achieve each of its many goods; or, if the good of a plant is also thought to be multiple – e.g. nourishment, growth, reproduction – all these things are a means to its one and most complete good, which is limited compared to the human good.

292b10-25 So one thing has and shares in the best, [another reaches it through a few <steps>, another through many, another does not even try for it, but it is sufficient <for it>[81] to come close to the ultimate.

292b13 For example, if health is the end, one person is always healthy, another thins to be healthy, another runs and thins, another performs some other action for the sake of running, so that its motions are more numerous, another is unable to be healthy, but only to run or thin (and one or the other of these things is the end for those people). It would be clearly best for each person to attain that end, but, if not, it would always be better to the degree that it got closer to the best. And that is why the earth does not move at all, and things close to it have few motions. For these things do not reach the ultimate, but they attain as much of the most divine starting point as they are able.[82] But the first heaven attains it straightaway with one motion. But the things in between what is first and what is last, do reach it,] but they reach it with more motions.[83]

Having said that one should not think about heavenly things as things without soul but as things having soul and engaging in action, and having presented the differences among things which engage in action, he comes to the issue set forth and provides the solution of the difficulty raised. As Alexander says, he uses what was said first in providing the explanation, that no action is necessary for what is in the best condition, other things need some slight action for attaining what is best, and others more. But perhaps he also mixes in the second determination which proves that a little motion is not always better, but is sometimes worse than more.[84]

And so he says that among existing things neither the first nor the last has need of action, the last because it does not directly attain the end, the first because it is not divided from the good but has it and shares in it in its own substance. The word 'has' might be said of the hypersubstantial goodness and the One, and the word 'shares' of the intellect which is directly unified with the good and shares in it; for the One is said to 'have' in its own substance because it projects something, intellect is said to share because it receives from something else. (That Aristotle has a conception of something above intellect and substance is clear at the end of the book on prayer where he says clearly that god is intellect or even something which transcends intellect.)[85]

Another thing reaches its own end through a few motions. But there are two kinds of end, one which is the best and most final of all things, the other more partial. And there are two senses of slight motion:[86] in one sense, the multitude of motions is brought together in itself and consequently it attains the common and whole end, in the other it is a part of many motions and consequently reaches towards a partial end. And it is clear that in the first sense a slight motion is better than many motions, but in the other it is worse, so that what attains an end through more actions and activities would be a mean among things which attain an end through slight motions.[87] In this way there is a solution of the difficulty why the sun and moon, which are further from the fixed sphere, which has one motion, have fewer motions than higher things, which are nearer to
486,1 the fixed sphere but have more. He says that this is because some things which have slight motions are better than things which have more motions and some are worse. And he has said which is which:

what is last does not even try to attain the goal immediately, so that it does not move either, but it is sufficient for it to come close to the end.

(292b13) Next he clarifies what he has said by means of the example of health (in the example, 'thinning' means 'becoming lean'); he says that attaining the most final end is best, but otherwise attaining what is as close as possible to that; and he next applies the present example to what he has said, moving from the last things to the first and then encompassing what is in between them. He says, 'that is why the earth does not move at all', its lack of motion not being for the same reason for which the good and that for the sake of which is motionless; the latter is that for the sake of which things which move move, and it is not necessary for it[88] to move. However, since the earth is last, it is not of such a nature as to share in that for the sake of which immediately, but to do so by being as near to things which share in it directly as is possible for it, and <in that way> sharing in it. The things close to the earth have few motions since they do not reach the ultimate end, i.e. the first and perfect good, because they are separated from it, but they move as far as they can in terms of sharing in the most divine starting point, and they can do this partially. And if he means by 'earth' earth in the strict sense, he would mean by what is 'close to' earth the elements which are above the earth and under the moon, but if he means by 'earth' everything sublunary, he would be calling the moon and the sun, which have few motions, close, and this is more appropriate to what he has said; for the difficulty concerned these things: why, when the fixed sphere has one motion, don't things further away from it, the sun and moon, always have more motions, whereas in fact the sun and moon have few motions, the things in between more? So, if, when he mentions few motions, he were not talking about the sun and moon, the solution of the difficulty would lack real authoritativeness; but if he is speaking about sun and moon, the words 'for these things do not reach the ultimate' seem to be harsh, unless he were saying that they are not made equal to its complete perfection because they are more partial; for he says clearly that these things share in the first starting 487,1
point according to their own measures, since he says that they share in the most divine starting point as much as they are able.[89]

He says that the first heaven attains the first starting point straightaway, i.e. immediately, with a motion which is one in form because this one motion contains, generates, and brings together all the motions. For the first heaven is the first thing moved, and it imitated through its perfect motion the perfection of what is motionless; and Plato[90] would say that what that perfection is with respect to what endures forever, the first heaven becomes with respect to all the time of the things under it which divide the sameness of that perfection. But if trusting in divine myths[91] is dear to someone, let him grasp from these things that the successor of the first heaven, the great Saturn, started the division and particularisation. But these matters are discussed elsewhere.

He speaks of 'the things in between what is first and what is last', meaning by what is first the fixed heaven, by what is last the sun and the moon. For these are the extremes of the divine body. He says that what are in between these things, being more universal than last things, do reach more towards the perfection of the starting point, but they reach it through a division of motions and not through one motion in the way the first heaven does, so that they reach it through several motions which divide up the one motion completely, and so they are said to reach to where the one motion also leads.

And it seems to me that Aristotle has considered all the heavenly motions which unfold the entirety of the unmoving unity and found the fixed sphere to have one motion which contains all the motions, the spheres after it to have all <motions> in a divided way, and the sun and moon not to have all motions; for these are not observed to have stations, backward motions, different *phaseis*, or retrogressions and progressions,[92] so that the astronomers, in providing explanations of the phenomena, are satisfied with simpler hypotheses <in their case>. And so, hypothesising that the motions are cases of action and occur on account of assimilation to the good, he says that the first heaven directly attains perfect assimilation to what is unmoving through one perfect motion, but the spheres after it reach perfect assimilation through all the divided motions, and the sun and the
488,1 moon, which do not have all the motions, share in it as much as possible.

In this way Aristotle has provided the solution of the difficulty, while making a concession to it and agreeing that the planets move with motions which are many in form because they do not just appear to move forward, they also appear to move backward and stand still, and they have different *phaseis* and retrogressions and progressions and all kinds of non-uniformities. In order to preserve these <apparent motions> the astronomers use many motions for each <planet>; some hypothesise eccentric <spheres> and epicycles, others homocentric <spheres> which are called counteractive. The true account does not accept that they stand still or move backward, or that there is addition or subtraction in the numbers of their motions[93] even if they are observed to move in this way, nor does it admit hypotheses of this kind; rather it demonstrates that the heavenly motions are simple, circular, uniform, and ordered, using as evidence their substance. Not being able to grasp with precision how what occurs in the heaven is only the appearance of their condition and not the truth, the <astronomers> were content to find out on what hypotheses the phenomena concerning the stars which are said to wander could be preserved by means of uniform, ordered, circular motions. And, as

Eudemus recorded in the second book of his astronomical history (and Sosigenes[94] took this over from Eudemus), Eudoxus of Cnidus is said to be the first of the Hellenes to have made use of such hypotheses, Plato (as Sosigenes says) having created this problem for those who had concerned themselves with these things: on what hypotheses of uniform and ordered motions could the phenomena concerning the motions of the planets be preserved?

If, then, the several motions for each of the several planets are hypotheses and they are not demonstrated to be this way in truth, as is made clear by the fact that different people make different hypotheses, what necessity is there to seek in this way the reason why the planets proximate to the fixed sphere have more motions than the last ones, as if there were in truth several bodies and therefore several motions in the cases of each of the planets?[95] Perhaps (if it is necessary for us to hazard making these kinds of comparisons at all), it is necessary to determine the worth of these things not with respect to the difference among their locations, but to say that each has been assigned to be where it benefits the universe. So, since sublunary 489,1
things do not have their own light, but are illuminated from outside, it is reasonable, one might say, that the two luminaries of the cosmos have been assigned to be directly above them, and perhaps they have simplicity in their motions because simplicity is better than compositeness.

Plato is thought to say in the *Laws* that the planets are observed to move in variegated ways, but do not move this way in truth, whereas in the *Timaeus* he agrees that their motion is more variegated because they are means between what is entirely ordered and what is entirely disordered, and therefore have an ordered non-uniformity. Consequently in the *Laws* he inveighs against those who predicate only wandering of them and do not consider that their wandering also participates in order and is natural for them.[96]

When Alexander is discussing this passage he says without further ado that the four sublunary elements are without soul and do not participate in action. But who would not be surprised if, when animals which are compounded out of very small portions of the elements[97] have soul (although their substance is short-lived and altogether contracted into a small span), such large portions of the universe, which are eternal in their entireties, were not judged worthy of soul by the demiurge? For even if they are simple, it would not be necessary that they be without soul, if the heaven, which is simple, was given soul,[98] and when each of the things which is compounded from the four elements is what it is called because of the predominance of one of the elements. But if they do not share in action, because they are not active in different ways at different times as particular animals are, the heaven, too, always has the same order of activities. But if <Alexander> thinks that the earth is without life and soul because it does not change place, first of all we ought to be ashamed if we say that plants which are given life by the earth live and have soul, but that the earth itself is without life and soul; and secondly, when Aristotle says that intellect and soul are alive,[99] he does not require that they move locally; and even if the earth, as the hearth of the universe,[100] stands still, it has this action and activity, since, just as moving in a vital way is action and activity involving soul, so is standing still in a vital way. And so heavenly things move, the earth stands still, and particular animals both move and stand still.

> **292b25-30** In the case of the difficulty [about there being one motion in the case of the first <sphere> although it is composed of a great multitude of stars, whereas each of the others has separately received its own motions, one might reasonably think that this holds primarily because of one thing; for one should think that, with regard to each's life and sovereignty,[101] there is a] great superiority of the first over the others ...

He has proposed two difficulties and solved the first. He now turns to
490,1 the second, which inquires why the fixed sphere, which is one and has one motion, has a multitude of stars so great as to be thought uncountable,[102] all having the single motion of the fixed sphere, whereas each of the stars which are said to wander has received its own motion from the sphere in which only it is. He solves this difficulty with two or three[103] arguments, the first of which he has taken[104] from the superiority which the fixed sphere has over the others; for even if all of them have life and a sovereign worth, one should grasp that there is a great superiority of the first over the others with regard to life and sovereignty. Its superiority in power is made clear by its immediate kinship with the first efficient and moving cause,[105] by the fact that it contains the others and carries them around with itself, and furthermore by the fact that it attains the most complete good by means of a single, simple, perfect motion which is, if one were to conceive its magnitude, practically instantaneous. Consequently one might rightly be more surprised by the opposite, that is, if, when it excels by so much, it nevertheless had some numerical ratio of its power to <that of> the other spheres, a ratio which the multitude of stars moved by it have to each single star fastened in the other spheres.

> **292b30-293a4** ... and this[106] will be a reasonable result, [since the first <motion>, which is one, causes many divine bodies to move, but each of the many others causes only one to move (for any one of the planets has several motions). So in this way

> nature equalises and makes a certain order, assigning many bodies to the one motion] and many motions to the one body.

They[107] say that one should not run this together with the preceding, connecting the two texts, but should understand it as a second argument. For he says that the first motion, that of the fixed stars, which is one, causes many divine bodies to move in accordance with its one motion, but the motions of the planets, of which there are several of several spheres in the case of each star, cause a single body to have several motions; for each of the planets has several motions, since it is moved by several of the so-called counteractive <spheres>. So, he says, in this way nature equalises the superiority which is so great and makes a certain order, assigning many bodies of the fixed stars to the one motion of the fixed sphere, and many motions to the one body of a planet.

The interpreters who count this argument as distinct in itself recommend that one not connect it with what precedes. But perhaps if it is not connected to what precedes, the words 'and this will be a reasonable result' are unintelligible. For what is 'this', if he is not referring to the superiority <of the fixed sphere>? To say, as Alexan- 491,1
der does, that 'this' is the thesis or the order or the apparent inequality makes the ellipsis of expression great. Perhaps then he is speaking of the great superiority of the fixed sphere over the planetary spheres and using it to solve the difficulty by proving next that this superiority is somehow equalised by the divine demiurgic creativity using proportionality. For as the one motion of the fixed sphere is to the many stars which are moved by it, so the one planet is to its many motions. For if what is said is not understood in this way, but it is treated as in itself a completely distinct argument, perhaps it also refutes what was said before,[108] since that makes superiority responsible, and this makes equalisation responsible.

> **293a4-11** And furthermore, the other spheres have a single body[109] [because the motions before the last one, which has a single star, move several bodies; for the last sphere which is moved is fastened in several spheres, but each sphere is a particular body. The task of that body will therefore be common, since, while each sphere itself has its proper natural motion, this motion is, as it were, added,] and the power of any finite body is related to a finite body.[110]

This is another argument (the second or the third)[111] providing, on the basis of the so-called counteractive spheres, an explanation of the fact that the planetary <spheres> have one star each and the fixed sphere has as many as it does. He says that the sphere which has a single star that is said to wander is moved, fastened in many spheres, which are called counteractive or (as Theophrastus calls them) starless;[112] this sphere is the last one in the whole system of spheres which cause, e.g., Saturn or Jupiter or some other <planet> to move. And there is a natural simple motion which is proper[113] to each of these spheres, both the one which has the star and those which contain that sphere, but the variegation and non-uniformity of the star, which seems to move forward and backward and to add to and subtract from its numbers[114] and to stand still, are added from outside <the sphere which has the star>; for they are produced by the counteractive spheres, since, as has been said, each of the spheres moves with its proper motion, and they each cause the sphere having the star to move in a different way in accordance with their own appropriate motion. So, since each sphere is a body, and in the case of the outermost sphere in each system, which moves with the fixed sphere, there is added to its own motion the fact that it also causes all the other spheres which it contains to move in common with the same motion with which it moves, it would be a difficult task for the outermost sphere to move both so many corporeal spheres and the sphere having the single star, if that sphere had many stars rather than one, as the fixed sphere does. Aristotle indicates the difficulty of the task when he says that 'the power of any finite body is related to a finite body'. For if what causes motion had infinite power, then nothing could overcome it with respect to being moved; but since it is a finite body and so has finite power, this power will be related to something finite and commensurable with it, and not to anything whatsoever. And if so great a number of bodies exceeds the kinetic power of the one sphere with respect to being moved by it, then, if the sphere having one star had many, the assigned task would be really difficult.

It seems to me that this argument goes forward <only> against the person who says that the fixed sphere has a great superiority over the planetary spheres. For what prevents it being the case that, just as the fixed sphere causes both all the stars in it and all the spheres contained by it to move along with it, so too the outermost counteractive sphere causes both the counteractive spheres under it and the one having the star, which <by assumption> no longer has one but many stars, to move? If the fixed sphere is not so superior, what is the difference between the fixed sphere (which exists along with its stars and has its own motion in such a way as to carry around the spheres which it contains) and the sphere having the planet (supposing it had a multitude of them) which makes motion caused by another a more difficult task for the <latter> cause of motion? (The difficulty of the task is not due to these bodies having weight, since Aristotle has denied this on the basis of demonstrations;[115] rather it is due to the fact that in the case of those things too there must always be a commensurability between what causes motion and what is moved. And Aristotle has based his demonstration <here> on this fact.)

One should note that this argument proceeds on the basis of the assumption that the astronomical hypotheses concerning the counteractive spheres truly hold, but, as I said previously,[116] there is no necessity to these hypotheses, since other people preserve the phenomena by means of other hypotheses. It would be appropriate to the discussion of the heaven and the heavenly motions to discuss briefly these hypotheses with which each of them strove to preserve the phenomena.[117] And I have also said previously[118] that Plato without hesitation assigned to the heavenly motions circularity, uniformity, 493,1
and order and put forward to the mathematicians this problem: by making what hypotheses about uniform, circular, and ordered motions will it be possible to preserve the phenomena involving the planets? And I have said that Eudoxus of Cnidus first proposed the hypotheses using the so-called counteractive spheres. Callippus of Cyzicus, who studied with Polemarchus, the associate of Eudoxus, went to Athens after him and lived with Aristotle,[119] and together with him corrected and filled out the discoveries of Eudoxus. Because it hypothesised that the counteractive spheres are homocentric (and not eccentric in the manner of later <astronomers>) the hypothesis of counteractive spheres appealed to Aristotle who thought that all the heavenly bodies should move around the centre of the universe.

It seemed to Eudoxus and his predecessors[120] that the sun has three motions: it is carried around from east to west by the sphere of the fixed stars; it moves in the opposite direction through the twelve signs of the zodiac; and, third, it turns to the sides of the middle of the signs of the zodiac (this was determined from the fact that it does not always rise at the same place at the summer and winter solstices).[121] Therefore he said that it is carried in three spheres – Theophrastus called these spheres starless because they have no star;[122] and they carry along what is lower and counteract what is higher. The sun has three motions, and it would be impossible for it to be moved in opposite directions by the same thing, at least if neither the sun nor the moon nor any other star moves on its own, but they are all fastened to and moved by a circular body. If, then, the sun were to complete its circuit in length and its departure in breadth <from the ecliptic> in one and the same time, two spheres would be sufficient; one, that of the fixed stars which makes a revolution towards the west; the other, turning towards the dawn around an axis fastened in the first sphere at right angles to the oblique circle – it is on this sphere that the sun would be thought to make its <west-east> progression. However, since this is not the way things are, but the sun goes around the one circle in one time and makes its departure in breadth in another, it is necessary to assume a third sphere in addition so that each motion will explain a different phenomenon

concerning the sun. So, in this way there are three spheres, all
494,1 homocentric with one another and with the universe; one sphere containing the other two was hypothesised to turn around the poles of the cosmos in the same direction as the sphere of the fixed stars and to be restored in the same time as it; a second sphere, smaller than this but greater than the third was hypothesised, as has been said, to turn from west to east around an axis which is at right angles to the plane through the middle of the signs of the zodiac; and the smallest was hypothesised to turn in the same direction as the second, but around a different axis which should be conceived as perpendicular to the plane of a certain great, oblique circle which the sun is thought to describe with its own central point when it is carried by the smallest sphere in which it is fastened. As is clear from his treatise *On Speeds*,[123] he <Eudoxus> posited that the retardation <produced by the third> sphere is much slower than that of the sphere containing it (the one which is intermediate in size and position). So the largest sphere makes both the remaining spheres turn in the same direction as the fixed stars due to the fact that the first carries in itself the poles <of the second> and the second carries in itself the poles of the third, which carries the sun, and similarly <the second> has in itself the poles on which it is carried around, turning along with itself both the third sphere and therefore at the same time the sun; in this way it results that the sun is observed to move from east to west. And if the two spheres, the middle one and the smallest, did not move on their own, the revolution of the sun would take the same amount of time as the revolution of the cosmos, but, in fact, since these spheres turn backward in the opposite direction, the return of the sun from one rising to the next lags behind the time of the revolution of the cosmos.

This is what was said about the sun. In the case of the moon <Eudoxus> ordered some things in the same way and some differently. He thought there are three spheres carrying the moon because it appeared to have three motions; one of these is the same as the motion of the fixed stars; a second turns in the opposite direction to this around an axis which is perpendicular to the plane of the <circle> through the middle[124] of the signs of the zodiac, as also in the case of the sun; but the third is no longer the same as in the case of the sun because, although it is similar in its position, it is not similar in its
495,1 motion; rather it moves in the direction opposite to the second, being carried in the same direction as the first in a slow motion and turning around an axis perpendicular to the plane of the circle which would be thought to be described by the central point of the moon and which is inclined to the <circle> through the middle of the signs of the zodiac to the extent of the greatest departure in breadth of the moon. But it is evident that the distance of the poles of the third sphere from those of the second would be the arc of a great circle conceived as passing

through the poles the length of which is half of the breadth through which the moon moves.[125] So, he hypothesised the first sphere because of the moon's circuit from east to west, the second because of its apparent retardation in the signs of the zodiac, the third because it is observed that it is not furthest north and furthest south in the same points of the zodiac but that these points in the signs of the zodiac always shift towards preceding signs.[126] Consequently, he assumed that this sphere moved in the same direction as the sphere of the fixed stars, but because in each month the shift in the points mentioned was very slight he assumed that its westward[127] motion was slow.[128]

He said this much about the moon. Aristotle sets out his opinion concerning the five planets.[129] He says that these are moved by four spheres, of which the first two are the same and have the same position as the first two in the case of sun and moon; for in the case of each of them the sphere which contains all <the other spheres> turns around the axis of the cosmos from east to west in the same time as the sphere of the fixed stars, and the second sphere, which has its poles in the first makes its revolution in the reverse direction from west to east around the axis and poles of the circle through the middle of the signs of the zodiac in the time in which each of them is thought to traverse the zodiac. Accordingly, he says that the second sphere completes <its circuit> in a year in the case of Mercury and the Morning Star,[130] in two years in the case of Mars, in twelve years in the case of Jupiter, and in thirty in the case of Saturn (which earlier people called the star of Helios).[131]

The last two spheres are arranged more or less in the following way. In each case the third sphere has its poles on the circle through the middle of the signs of the zodiac, the circle which is conceived of 496,1 as in the second sphere in each <system>,[132] and it turns from south to north,[133] going through all its relations to the sun, in the same time in which its planet moves from one *phasis*[134] to the next – the mathematicians also call this the time of traversal.[135] This time is different for different <planets>, so that the revolution of the third sphere does not take the same amount of time for all of them; rather, according to what Eudoxus thought, it takes nineteen months in the case of Venus, one hundred and ten days[136] in the case of Mercury, eight months and twenty days in the case of Mars, and very close to thirteen months in the case of each of Jupiter and Saturn.[137]

So the third sphere moves in this way in this much time. The fourth, which also carries the star, turns in a certain oblique circle with, in each case, its own poles; it completes a revolution in the same time as the third sphere but it moves in the opposite direction from east to west.[138] He <Eudoxus> says that this oblique circle is inclined relative to the greatest of the parallel circles in the third sphere,[139] but not inclined equally or in the same way in all cases. It is, then, evident that the one sphere which turns in same way as the fixed stars makes all the remaining spheres turn in the same direction because each has its poles in another, and accordingly does so in the case of the sphere carrying the star and in the case of the star itself. And for this reason each of the planets will rise and set. And the second sphere will make its passage under the twelve signs of the zodiac, since it turns around the poles through the middle of the signs of the zodiac and makes the remaining two spheres and the star turn with it towards successive signs in the time in which the star is thought to complete the zodiacal circle. The third sphere has its poles on the circle through the middle of the signs of the zodiac in the second sphere; it turns from south to north and from north to south, and it will make the fourth sphere, the one containing the star, turn with it, and it will contain the cause of the star's motion in breadth. However, not it alone, since to the extent that the star moved on this circle it would reach the poles of the circle through the middle of the signs of the zodiac, and it would also come to be near to the poles of the cosmos. But now the fourth sphere, which turns around the poles of the oblique circle of the star from east to west in the opposite direction to the third and makes a revolution in an equal time as it, will prevent an excessive departure from the circle through the middle of the signs of the zodiac and will make the star describe what Eudoxus called a 'horse fetter'[140] about this same circle.[141] Consequently, the star will seem to depart in breadth from the ecliptic by the amount of the breadth of this line <the horse fetter>. People attack Eudoxus for this.

Eudoxus' description of the sphere uses twenty-six spheres in all for the seven planets, six for sun and moon and twenty for the five <other planets>. Aristotle has written this about Callippus in Book 12 of the *Metaphysics*:

> Callippus posited the same position for the spheres, i.e., the order of their distances,[142] as Eudoxus, and he assigned the same number of spheres to Jupiter and Saturn as Eudoxus, but he thought that two spheres had to be added to <each of> the sun and moon, if one was going[143] to explain the phenomena, and one more[144] to each of the other planets.[145]

So, according to Callippus, there are in all five times five and two times four, that is thirty-three, spheres.[146] There does not survive a treatise by Callippus stating the reason that these additional spheres must be added, nor does Aristotle supply the reason. But Eudemus recounts briefly for the sake of which phenomena <Callippus> thought that these spheres had to be added. He says that <Callippus> says that since the times between the solstices and the equinoxes differ by as much as Euctemon and Meton thought,[147] the three spheres for each <of sun and moon> were not sufficient to preserve the phenomena, obviously because of the apparent non-uniformity in their motions. And Eudemus also recounts briefly and clearly for what reason <Callippus> added one sphere in the case of each of the three planets Mars, Venus, and Mercury.[148]

After recounting Callippus' opinion about the counteractive spheres, Aristotle adds:

> But if all the spheres added together are going to explain the phenomena, it is necessary that there be for each of the planets other spheres, one less <than the number of spheres associated with the planet>, which are counteractive and always restore the first sphere of the star beneath it in order to the same position;[149] for only in this way can they all produce the motion of the planets.[150]

So Aristotle has stated these things briefly in this way and[151] clearly; 498,1
Sosigenes praises him for his acumen and tries to find the use of the spheres added by him; he says:[152]

> It is necessary for these spheres (which <Aristotle> calls counteractive) to be attached to the hypotheses for two reasons: so that there will be the proper position for both the fixed sphere for each planet and for the spheres under it; and so that the proper speed will be present in all of the spheres. For it was necessary both that a sphere move in the same way as the sphere of the fixed stars or as some other sphere around the same axis as it and that it rotate in an equal time, but neither <property> could possibly belong to it without the addition of the spheres mentioned by Aristotle. For the sake of clarity let us give the explanation in terms of the spheres which move Jupiter. If we were to fit the poles of the first sphere for Jupiter into the last of the four spheres for Saturn, the one in which Saturn is fastened, how could the poles remain on the axis of the sphere of the fixed stars when the sphere carrying them is turning around another axis which is to the side? However, in the case of the outermost motion it is necessary for them to remain on that axis if the sphere which turns around them is going to receive the ordering of the sphere of the fixed stars. Furthermore, since the three spheres carrying Saturn are turned by one another and the first sphere, and they have some speed of their own, the fourth sphere will not have some simple motion, but a motion which shares in all those above it. For it will be shown[153] that something of the speed belonging to spheres moving in the contrary direction is subtracted from them by the sphere which causes turning in the same direction and that in the case of those turning in the same direction something is added to the

motion which penetrates to them from the spheres above because of the motion of those spheres. Consequently, if the first of the spheres of Jupiter were fitted into the sphere carrying Saturn and the first sphere of Jupiter had its own speed which would make it return again to the same place in a rotation of the cosmos, the motion of the spheres above would not permit it to have its present speed, but there would be some addition, since they are moved towards the west, and the first sphere of Jupiter would itself be moving in the same direction.

The same argument also applies to the next spheres. For the
499,1 motion will be more and more compounded and their poles will change from their proper position. But it is necessary, as we said, that neither of these things happen.[154] In order that this not occur and that one encounter nothing unsatisfactory because of it, he <Aristotle> conceives of '<spheres> which are counteractive and which always restore the first sphere of the star beneath it in order to the same position.'[155] For he says exactly these words and he is referring to both of the reasons for which he introduces these things. When he says 'counteractive', he is referring to the restoration of the motion to its proper speed, and when he says 'which always restore the first sphere of the star beneath it in order to the same position', he is referring to the poles remaining in the suitable place. For the position of the moving spheres is grasped by reference to the poles only if they remain fixed.[156] But he says that the first sphere of the star beneath <another planet> in order is restored by the <counteractive spheres of the star above> since <the first sphere> receives its appropriate speed and appropriate position from the counteraction, and all the speeds and positions of the next spheres are preserved.

Sosigenes proves that these things result, after first setting out some things which are serviceable for the argument. What he sets out is, in brief,[157] this:

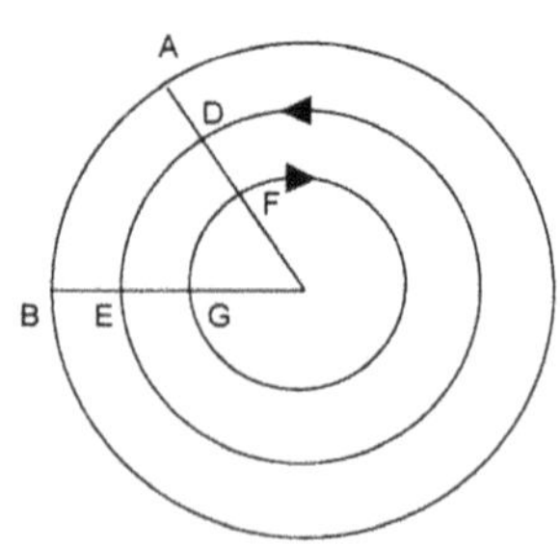

Figure 1 (499,16-500,14)[158]

Suppose there are two homocentric spheres, DE, FG, which are contained by another outside them, the one containing them being either in motion or remaining fixed;[159] and suppose they move in directions opposite to each other and in an equal time, i.e. at the same speed; then all the points in the contained sphere will always be in the same position relative to the containing sphere, as if the points in fact remained fixed.[160]

For suppose DE is moving from A towards B.

If the smaller sphere FG were only turning with DE and not moving in the contrary direction, it is seen that just as D is at some time under B, so F moves together with D and in the same time. But since they also move together and FG also moves in the contrary direction to DE, however much <FG> adds by moving in the same direction, it cancels by moving in the contrary direction. The result is that when D is under B, F is under A, just as F was observed to be originally. Consequently the proposition is true. So, if AB remains fixed what has been proved is clear and it is clear that[161] if both things hold and the inner sphere <FG> moves around with the outer one <DE> and also moves in the contrary direction, then the same points will always have the same position, but not if <FG> moves only in the same direction as <DE> or only in the contrary direction.

But if the sphere AB also were to move, either in the opposite direction or in the same direction as the second sphere DE, the same thing would happen in the case of the points on the third sphere FG, which moves together with DE and moves in the contrary direction at an equal speed. For if the sphere AB turned from A towards B and drew DE with it so that D moved to E, the middle sphere itself, DE, would move either in the opposite direction or in the same direction as AB and at some speed or other relative to AB and in an equal time with FG; and because <DE> makes the third sphere <FG> turn in the same direction as it, it makes the point F go beyond A. But the third sphere moving in the contrary direction will make F again be under A, and, since this will always happen, all the points on the sphere FG will be under the same points on the sphere AB.[162]

So the proposition has been proved in the case in which the spheres move around the same axis, and the argument is the same even if they do not move around the same axis. For the position of the points under the same points does not result because their motion is along the same parallels, but because the contained sphere is carried around by the containing sphere and also moves in the contrary direction, and however much it adds it subtracts, whether the carrying around in the two directions occurs in the case of an oblique circle or an upright one.

Again, suppose there are two homocentric spheres moving in the same direction, each with some speed, the smaller not only being carried together with the greater but also having its own motion in the same direction. Then, if their speeds are equal, <Sosigenes> will prove that the compound motion is double in speed; but if the speed of one is double <that of the other>, the speed of the compound will be triple, and so on. For if the greater causes the smaller to move a quarter circle and the smaller moves at the same speed, then the smaller itself also moves a quarter circle, and a quarter circle will have been moved through twice, so that the motion compounded of both is double the motion of the one. <Sosigenes> says:

> 501,1 We say these things if the motions are around the same poles. But if they are not, something else will happen because of the obliqueness of one of the spheres. For then the speeds will not be compounded in this way, but in the way it is usually proved in the case of a parallelogram where the motion along the diagonal is produced from two motions, one of a point moving on the length of the parallelogram, and one of this length itself drawn down in the same time through the breadth of the parallelogram. For the point and the side of the length which has been drawn down will be together at the other end of the diagonal; and the diagonal will not be equal to <the sum of> both of the lines which are broken at it, but it will be less, so that also the speed of the compound will be less <than the sum of the two speeds> although it is compounded of the two.[163]

And <Sosigenes> says the following in a way quite similar to what he has said:

> If there are two homocentric spheres, whether having the same or different poles, being carried around in opposite directions with the smaller being carried around by the greater and also moving at a lesser speed in the contrary direction, the points in the smaller sphere will reach the same place in a longer time than if the smaller sphere were only fastened in the greater <and not moving in the contrary direction>. Because of this fact the restoration of the sun itself from rising to rising is slower than the rotation of the cosmos because it moves more slowly than the universe and in the opposite direction, since if it moved in an equal time to the fixed sphere and it were carried around in the contrary direction and always were restored in the same time, it would always rise together with the same point.

These things being taken to start, Sosigenes comes to the things said by Aristotle concerning the necessity for other counteractive spheres for all but one of each of the planetary spheres, if the phenomena are going to be explained. And he sets out the theory of Aristotle's description of the sphere as follows:

> The first of the spheres carrying Saturn moved in accordance with the sphere of the fixed stars, the second moved in the circle through the middle of the signs of the zodiac and was left behind <by the first>, and the third moved from south to north through the circle perpendicular to the circle through the middle of the signs of the zodiac, which it carried aside in breadth – this circle was perpendicular to the circle through the middle because it had its poles on that circle, and circles which cut through the poles, cut perpendicularly.[164] The fourth sphere, which has the star, moved the star on a certain oblique circle which delimits the breadth of the planet's turning away to the north so that it does not come near the poles of the cosmos.
>
> And then one should conceive an additional fifth sphere, prior to those carrying Jupiter,[165] moving around the same poles as the fourth, in the opposite direction as it, and making a revolution in the same time; this sphere will cancel the motion of the fourth sphere because it moves around the same poles as it, but in an opposite direction, and in the same time – for this has been proved – and it will diminish the apparent speed.[166] After the fifth, in order to preserve the phenomena, one should conceive another sixth sphere with the same poles as the third, which it counteracts, and moving in the opposite direction in the same time; and the points on the third sphere will always appear directly below on the sixth sphere. And after this sphere one should add a seventh, counteracting the second, which is fitted in around the poles of the circle through the middle <of the signs of the zodiac>, around which the seventh also moves, but turning in the opposite direction to the second and in the same time as it. The seventh sphere cancels the motion and the speed which penetrates through from the second sphere to the spheres beneath it; for the seventh sphere, moving along with the fixed sphere, adds to the speed <of the motion> from east to west of the spheres under it.[167] So it will turn in this way and move in the same way as the fixed sphere, but it will not have the order of the fixed sphere, since it turns around poles different from those of the fixed sphere; but it does move from east to west.

After the seventh sphere one should conceive a final eighth sphere, the first of Jupiter. Sosigenes correctly notes that the last of the three counteractive spheres <for Saturn> is not, as some people think, the first sphere of Jupiter, because then the last of the spheres counteracting the higher motions would be the first of the spheres moving the

planet beneath so that the seventh sphere and what we call the eighth would be the same, and be the first sphere of Jupiter; for it turns out for those people who try to preserve the number of counteractive spheres stated by Aristotle that they count the same sphere twice. For, in the case of each star the counteractive spheres should be one less than those moving the sphere; consequently there being four spheres moving each of Saturn and Jupiter, there will be three counteractive spheres, and there being five spheres moving each of the four, Mars, Venus, Mercury, and the sun, there will be four. So
503,1 the total of counteractive spheres is two times three for Saturn and Jupiter and four times four for Mars, Venus, Mercury, and the sun, and the total is twenty-two. But there were eight spheres carrying Saturn and Jupiter, and twenty-five carrying the remaining five. So if these thirty-three spheres are added to the twenty-two counteractive spheres, the total will be fifty-five. (For no spheres are needed to counteract those moving the moon, since it is last and Aristotle says that only the spheres in which the star lowest in order is moved need not be counteracted.)[168]

It is clear that this is the number of all the spheres. But Aristotle says that if one does not add to the sun and the moon the motions we have spoken about, the total will be forty-seven, and this remark produces consternation. For if we subtract the two spheres of the sun and moon which Callippus added and obviously also two others from the sun, namely the ones counteracting the two added spheres (since if those two are subtracted, one should also subtract the spheres intended to counteract them), six spheres will be subtracted, two carrying the sun and two counteracting them in addition to the two added by Callippus for the moon. But when these are subtracted from the fifty-five it will not result that there remain forty-seven in all, but forty-nine. But Aristotle says that forty-seven are left, perhaps because he has forgotten that he has subtracted only two, not four, spheres for the moon, unless one should say that he has subtracted the four counteractive spheres which he added for the sun and the spheres for both sun and moon which Callippus added, so since eight spheres are subtracted from the fifty-five, the remainder is forty-seven. The number <given by Aristotle> results in this way, but we will not be able to say why there are no spheres counteracting the two spheres of the sun, the second and third, but <Aristotle> says that only the <system> in the lowest position is not counteracted.[169]

However, Sosigenes also correctly notes that it is necessary to hypothesise counteractive spheres for the moon[170] in order that the speed added from the spheres above it to those moving the moon do not make it depart towards the west at a speed which is not the same as the fixed sphere. But given that only <the moon> does not have a counteractive sphere,[171] the number <47> does not result, and this caused both Alexander[172] and Porphyry in his commentary on Book

12 of the *Metaphysics*[173] consternation.[174] Sosigenes notes this and says that it is better to believe that the mistake in the number has come from the scribes[175] than to make the seventh and eighth spheres identical;[176] nor, if this were done, would the number harmonise with <Aristotle's> words since there will not then still be fifty-five spheres in all, as he himself says.

Sosigenes correctly[177] adds the following:

> However, it is clear from what has been said that Aristotle has one thing in mind when he calls these spheres counteractive, Theophrastus another when he calls them restorative.[178] They have both <features>. For they counteract motions of things above them and restore the poles of the spheres under them; they subtract motions, and they set motions in the right way. For it is necessary that motions not penetrate from above to the motions[179] of lower stars and that the poles of lower spheres fall directly below the poles of similar spheres so that the first spheres of stars which are lower in the ordering are restored to the same position (and obviously because of the first spheres the lower ones as well). For only in this way is it possible for all of them to make the motion of the fixed stars, as we have already said.

The description of the sphere by means of counteractive spheres is something like this, and it is not able to preserve the phenomena. Sosigenes also makes this accusation when he says:[180]

> However, the <hypotheses> of the associates of Eudoxus do not preserve the phenomena, and not just those which were apprehended later, but also those which had been known previously and were accepted by them themselves. And what necessity is there to speak about other things, some of which Callippus of Cyzicus also tried to preserve when Eudoxus had not been able to do so, whether or not Callippus did preserve them? But this very thing, which is also manifest to sight, none of them until Autolycus of Pitane[181] proposed to validate through hypotheses (and even Autolycus himself was not able to establish it). This is made clear by his disagreement with Aristotheros.[182] What I mean is that there are times when the planets appear near, but there are times when they appear to have moved away from us. And in the case of some <planets> this is apparent to sight. For the star which is called Venus and also the one which is called Mars appear many times larger when they are in the middle of their retrogressions so that in moonless nights Venus causes shadows to be cast by bodies;[183] and also in the case of the moon it is apparent to sight that it is not always equally distant from

us because it does not always appear to us to have an equal magnitude, although the conditions under which it is seen are the same.[184] And indeed the same thing is accepted by those who make observations using instruments because sometimes a seven-inch disk and sometimes an eight-inch[185] disk, placed at an equal distance from the observer, blocks his vision so that <the moon> does not fall on it. In addition to these things what occurs in the case of total eclipses of the sun bears witness to what has been said and is evidence for the truth about this question. For when the central point of the sun and the central point of the moon and our eye happen to be in a straight line, what occurs does not always appear similar; but sometimes the sun itself is also contained by the cone containing the moon and having our eye as vertex, and then it remains invisible to us for a certain time; and sometimes it deviates from this by so much that there is left at the temporal midpoint of the eclipse a kind of rim of the sun visible outside <the moon>.[186] Consequently it would be necessary that the <heavenly bodies> appear to be different in size when atmospheric conditions are quite similar because of the inequalities of the distances.

What occurs in these cases <of variations in distance> is also clear to sight, and it is reasonable that it also occurs with the other planets even if it is not manifest to sight; and it is not just reasonable, it is also true, since their motion each day appears non-uniform. But in the case of the apparent sizes <of these planets> there is no striking difference because their change upward and downward[187] (which the mathematicians customarily call their motion in depth)[188] also does not involve much difference. Consequently <these people> did not try at all to preserve this <phenomenon>, so that they did not show that the motion changes each day, although the problem[189] requires this. However, it is possible to say that the inequality of the distances of each <planet> in relation to itself did not escape them. For Polemarchus of Cyzicus[190] clearly recognises it, but, on the grounds that it is not perceptible, he ignores it because he prefers that the position of the spheres themselves in the universe be around the centre itself. Aristotle also makes this clear in his *Physical Problems* when he introduces difficulties for the hypotheses of the astronomers on the basis of the fact that magnitudes of the planets do not appear equal.[191] So he is not entirely satisfied with the counteractive spheres, even if he[192] does propose that the spheres are homocentric with the universe and move about its centre.[193]

And it is also evident from what <Aristotle> says in Book 12 of the *Metaphysics* that he does not think that the facts concerning the motions of the planets are sufficiently described by the astronomers before and during his time. At least he says the following sort of thing:

> In order[194] to get some conception <about the number of the heavenly motions> we will state what some mathematicians say in order to have some determinate number for our mind to take hold of,[195] but as far as the other things are concerned we should investigate some ourselves and learn others from[196] those who 506,1
> investigate; and if some of those who concern themselves with these things think differently from what we say now, we should love both, but listen to the more precise.[197]

But also in the same book, after counting all the motions, he adds:

> Let[198] the number of motions[199] be this great; and so it is reasonable to take there to be equally many substances and unmoving and perceptible[200] principles. Let us leave it for stronger <minds> to talk about necessity.[201]

The words 'let' and 'reasonable' and 'leave it for stronger <minds>' indicate his uncertainty on these matters. So, listening to Aristotle, one should rather follow those who came after him on the grounds that they preserve more of the phenomena; and even if these later people do not completely preserve them, those earlier people did not know as many phenomena, since the observations sent back from Babylon by Callisthenes,[202] when Aristotle requested them from him, had not yet reached Greece (Porphyry recounts that these observations were preserved for thirty-one thousand[203] years until the time of Alexander of Macedon); nor were those earlier people able to validate as much as they did know by means of hypotheses.[204]

Ptolemy[205] also censures these people because they introduce such a great multitude of spheres only for the sake of the simultaneous restoration of the seven planets in relation to the revolution of the fixed sphere[206] and for saying that contained and last spheres are responsible for the simultaneous restoration of containing ones above them, even though nature always makes the higher things responsible for the motion of lower ones[207] (since in our case impulses to move are communicated through the nerves from the hegemonic part to all the organs).[208]

But I do not know why for each star they arrange the first sphere in the same way as the fixed sphere and have it move at the same speed and they restore all the spheres after it down to the one having the star simultaneously with the fixed sphere. For if the sphere which lies above communicates the form of its own motion to the lower ones, why should we not say that the fixed sphere is the strongest and most dominant of all the spheres and restores all those under it simultane-

ously with it? For, since, in the case of each star the spheres which cause motion in length and in breadth are different, it was necessary that the spheres restoring them be different. But since the simultaneous
507,1 restoration together with the fixed sphere is the same for all the spheres, why is he not satisfied with <having> the planets carried around by the fixed sphere rather than <having> each star have need of spheres producing this motion and – according to Aristotle – spheres counteracting them? He would perhaps say that even if they are restored simultaneously with the fixed sphere because they move from the east with the same motion as it, nevertheless, because they differ in size they also differ completely in the speed of their motion; so how would it be reasonable that, when they have been severed and not bound together, they have their different motions because of the one fixed sphere?

So, in censuring the hypothesis of counteractive spheres, the people who came later deprecated the homocentric counteractive spheres most of all because they did not preserve the difference in depth and the non-uniformity of motions; and they hypothesised eccentric circles and epicycles – unless the hypothesis of eccentric circles was conceived by the Pythagoreans, as some others, Nicomachus and, following him, Iamblichus, recount.[209] In order that in writing a treatise on the heaven, we get some conception of the use of these hypotheses, let us first set out, in terms of a diagram, the hypothesis of eccentricity by comparison with that of homocentricity.

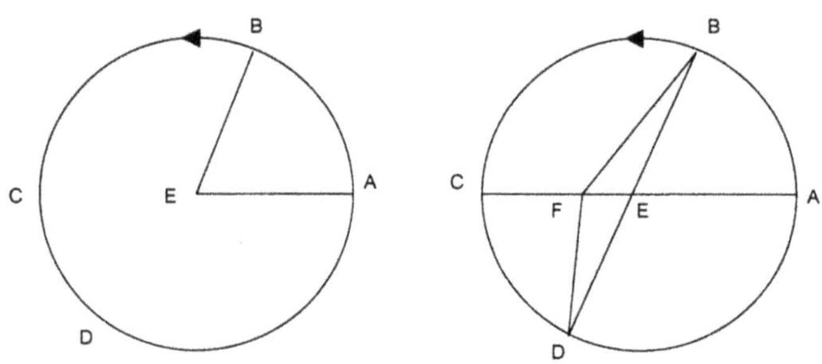

Figure 3 (507,18-25; 507,27-508,16)

Let the homocentric circle ABCD with central point E through the middle of the signs of the zodiac be conceived; let it be assumed that our eye is at E, and let AEC be a diameter. If a star makes a uniform passage from A to B on the circle ABCD, it is evident that, since our eye is at the central point E, if we conceive the ray from it and falling on the star as the straight line AE, it too will be carried around uniformly and obviously the star will also be observed making a passage uniformly and always at the same distance from us. However, since this is not the way they are in fact seen to behave, but rather they are seen to make their passage non-uniformly and to be at different distances at different times – as is made clear by the differences in their <apparent> sizes – let the circle ABCD be assumed to be not homocentric with the zodiac in the sense that it is no longer the central point of the zodiac E on which we say the eye is, but F, and ABCD is no longer homocentric with the circle through the middle of the signs of the zodiac but eccentric relative to it;[210] <and let> its apogee, that is, its greatest distance from the eye at F, be A, its perigee, its least distance from the eye at F, be C. If, then, we conceive the star moving in the same way on the eccentric circle ABCD and uniformly traversing the arc AB from the apogee A to B, and also a straight line from the central point[211] <E> being carried around with it, it will also move around uniformly. Let this line be EB. Then, if FB is joined from the eye at F to the star, it will result that the star has moved uniformly through the angle AEB but appears to have moved through the smaller angle AFB; for, since the angle <AEB> at E is exterior to the triangle BEF, it is greater than the interior and opposite angle <BFA> at F.[212] But if, in making its passage from the perigee C, it moves uniformly through the arc CD (so that the straight line ED is also carried around uniformly), and we again join the straight line FD from the eye at F, the uniform passage from the perigee will again be contained by the angle CED, the non-uniform apparent passage by CFD, and obviously its apparent passage <seen> from F[213] will be greater than its uniform passage because the angle <CFD> at F is greater than the <angle CED> at E.[214] And the angle AEB in the case where the star's position is B will be uniform,[215] but the angle AFB will be apparent, and the difference is EBF;[216] but in the case where the star's position is D, the angle CED will be uniform, CFD apparent, and the difference is EDF.

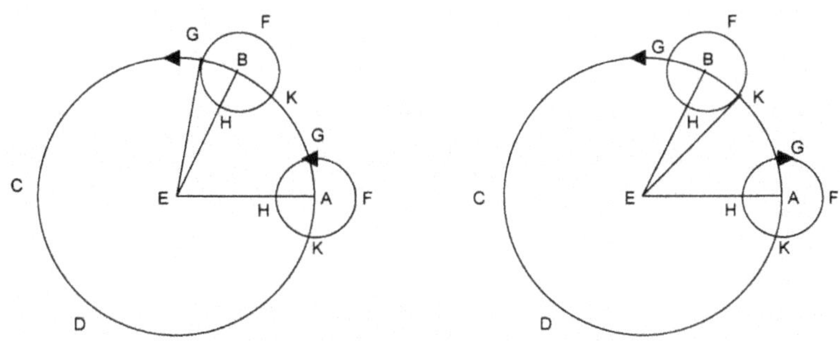

Figure 4 (508,22-509,3; 509,3-6)

In terms of being simpler this hypothesis harmonises with the stated purpose of the mathematician, and they have found another hypothesis which can validate the same things as the preceding one,

that is, make it that the stars are observed to traverse arcs of the circle through the middle of the signs of the zodiac non-uniformly although they are moving uniformly. For, again, let the circle ABCD with central point E, at which again our eye is, be conceived homocentric with[217] the circle through the middle of the zodiac; and <let> the star not move on this circle but on the circular FGHK, called an epicycle, which always has its central point A on the circumference of the circle ABCD; and let it move in such a way that the star reaches its apogee at F, in the same way as before, and its perigee at H. And it is clear that, when the epicycle moves uniformly through the arc AB, it comes to B, and the straight line EB is again carried around uniformly; but the star makes its passage from the apogee F to G, again moving uniformly through FG. And we will join the straight
509,1 line EG from the eye at E, and again the star will be moved around uniformly through the arc AB by the epicycle, i.e., through the angle AEB. But it will appear to move through the angle AEG, which is greater than the uniform angle <AEB>, the angle BEG being the difference between them. But when it makes its passage from the apogee not to G but to K, again the angle AEB will belong to the uniform passage, the angle AEK to the appparent one, which is smaller than the uniform one, the difference between them being KEB. So a hypothesis of this kind can validate that the passages of the stars towards their apogees is both greater and smaller, greater, obviously, when the star makes its passage from the apogee of the epicycle in the same direction as the circle, smaller when it makes its passage in the opposite direction; but the eccentric passage towards apogee always appears less than the uniform one since the apparent angle AFB is always smaller than the uniform angle AEB.

Each of these hypotheses taken by itself satisfies the purpose of astronomy, except that in the case of the moon they require both additions. For in order that the phenomena be preserved by the hypothesis, they hypothesise that the epicycle carrying the moon is carried around on an eccentric circle.[218] These hypotheses are also simpler than previous ones since they do not require inventing so many heavenly bodies and they preserve the phenomena, both the other phenomena and especially those related to depth and non-uniformity. However, they do not retain the axiom of Aristotle according to which every body which moves in a circle moves around the centre of the universe, but neither do they leave room for the solution of the difficulty which has been stated[219] and which has given rise to all these discussions. For there is no room left for equalisation since it is no longer true to say that the first motion, which is single, causes many divine bodies to move, and the motions which are many cause only one body to move; for the motions before the last, which has the single star, do not move many bodies.[220] Sosigenes also adduces these anomalies against these hypotheses, but he is not satisfied with the hypothesis of counteractive spheres for the reasons previously stated.[221]

But it is also necessary for those who believe that the stars have their own motion because they have soul to object to the first <axiom of Aristotle>, since each star is not just a part of the heaven, it is also a whole in itself. So a true axiom would instead be one which says that every body which moves in a circle moves about its own central point, so that it is true to say that whatever heavenly body has the centre of the universe as its centre moves around the centre of the universe, but whatever body, being more particular, is outside that 510,1
centre, moves about its own centre, as the stars[222] and epicycles and eccentric circles (if there are such bodies in heaven) do. But these things move about the centre of the universe, if not with their own motion, then with that of the sphere which is homocentric with the universe and carries them.[223] And for this reason Aristotle's doctrine that every body which moves in a circle moves about the centre of the universe would also be true, unless one were to add that it is moving with its own motion.

There will be partial room for the solution of the difficulty even on these hypotheses <of eccentric spheres and epicycles>. For on these hypotheses it is in a way true to say that nature equalises and produces a certain order by assigning many bodies to one motion and many motions to one body. For even if each has its own single motion, nevertheless everything under the fixed sphere will have its motion: the epicycles will have this motion plus the motion of the homocentric or eccentric circles; and the star, which he calls one body,[224] will have the motion of the epicycle and the motion of the homocentric or eccentric circles and the motion of the fixed sphere.

Furthermore, the eccentric circles would not be moving in a circle since they do not move around the centre but about what is outside the centre; and, since in turning they take up a place and leave it behind, they make it necessary for there to be a void; and their shape will be strange since what is inside is always cutting off a part of what is outside. Perhaps we will escape all these problems if we fit eccentric spheres into homocentric ones and say that a homocentric sphere moves around its own central point and carries the eccentric one (which also moves around its own central point) around with it. And we will say that they are all perfect spheres, not fearing to say that in their case a body passes through a body.[225]

Sosigenes also raises quite a few other astronomical difficulties for these hypotheses in an excellent way; these should be considered on another occasion when there is time. Now it would seem that, having investigated the theories concerning heaven and heavenly motions and confirmed the demonstrations by which they are proved to be circular, uniform, and ordered – since they are observed to be non-uniform and to have upward and downward motions – he has provided a conception of the hypotheses on the basis of which the earlier and later astronomers preserve the phenomena by means of uniform, circular, ordered motions. Now if this is more suitable for the discussion of the heaven than for that of first philosophy, no one will accuse us of turning the discussion aside for too long if the digression has been opportune. But we should pass on to Aristotle's next discussion.

[Chapter 13]

511,1 **293a15-27** It remains to speak about the earth, [where[226] it lies, whether it is stationary or moving, and about its shape.

293a17 Not everyone has the same opinion about its position. But most people say it lies at the centre <of the cosmos> (indeed, everyone who says that the whole heaven is finite does so). However, the Italians who are called Pythagoreans say the opposite, since they say there is fire at the centre and that the earth, which is one of the stars, moves in a circle around the centre to make night and day. Furthermore, they contrive another earth opposite to this, which they call by the name of 'counterearth', but they do not seek theories and explanations by relation to the phenomena; rather they force the phenomena to fit certain of their own theories and opinions] and try to put them in order together.[227]

He turns to the tenth and last topic of the book,[228] the earth, a topic which is also appropriate to discussion of the heaven. For he referred previously to the earth, saying that it lies at the centre of the whole heaven,[229] is motionless,[230] and has the relation of central point to it;[231] and he demonstrated the sphericity of the heaven from that of the earth.[232] Now he demonstrates in an unqualified way the things he assumed hypothetically about the relation of the earth to the heaven so that his discussion of the heaven will be complete. For the astronomers who are concerned with heaven and heavenly things also prove these things about the earth – that the earth is at the centre, that it is motionless, that it has the relation of central point to the heaven, and that its shape is spherical, since all these things belong to the earth in its relation to the heaven; and he will describe all the other things which belong to the earth because of its coordination with the other three elements in the two books which follow.

Now he proposes three issues about the earth: about its position, where it lies, and second 'whether it is stationary or moving', and third 'about its shape'. He first sets out the opinions on these subjects which have been put forward earlier, responds to them, and then argues for his own opinions on the subjects.

(293a17) And first he says about the position of the earth that those who say that the cosmos is infinite would not inquire about its position in the universe since there is no starting point, middle, or end in what is infinite. Most of those who say that the cosmos is finite, e.g., Empedocles, Anaximander, Anaximenes, Anaxagoras, Democritus, and Plato, say that the earth lies at the centre.[233] But the Pythagoreans contradict this (this is what is meant by 'say the opposite'); they say that the earth is not around the centre, but that there is fire in the centre of the universe. And they say that the counterearth, which is an earth, moves around the centre (they call it the counterearth because it lies opposite to this earth), and they say that the earth comes after the counterearth and it, too, moves around the centre and the moon comes after the earth (this is what he recounts in his work on the Pythagoreans).[234] The earth, being one of the stars and moving around the centre, makes night and day because of its relation to the sun; the counterearth moves around the centre and follows this earth, but it is not seen by us because the body of the earth always stands in front of us. <Aristotle> says that they say 512,1 these things because they do not seek theories and explanations which harmonise with the clear facts; rather they force the phenomena to fit certain of their own opinions and theories and try to make the apparent facts harmonise together with them – and this is very strange. For since they hypothesised that ten is the perfect number, they wished also to bring the number of bodies moving in a circle to ten. And <Aristotle> says that, assuming the one fixed sphere, the seven planetary spheres, and this earth, they filled out the ten with the counterearth.[235]

This, then, is the way he understands the Pythagorean doctrines. But those who share in them in a more genuine way say that the fire in the centre is the demiurgic power which generates living things from the centre of the whole earth and heats its parts which have grown cold. This is why some of them call fire the tower of Zeus,[236] as <Aristotle> recounts in his *Pythagorica*, others the guardpost of Zeus,[237] as he says in this work,[238] and others the throne of Zeus, as others say.[239] They called the earth a star on the grounds that it too is an instrument of time, since it is the cause of day and night – it makes day when the part facing the sun is illuminated, night because of the cone which is produced from its shadow. And the Pythagoreans called the moon the counterearth and also aitherial earth both on the grounds that it blocks the solar light, which is a special feature of earth, and on the grounds that it is the limit of heavenly things, just as the earth is of what is sublunary.

293a27-b4 Many others might also accept [that one should not assign the earth the space at the centre, but they do not found their belief on the phenomena but rather on arguments. For they think that it is suitable for the most honourable space to belong to the most honourable thing, that fire is more honourable than earth, that the limiting is more honourable than the intermediate, and that the extremity and the centre are limits; reckoning on the basis of these ideas, they think that earth does not lie at the centre of the sphere, but rather fire does.

293b1 Furthermore, because it is most suitable that the most authoritative <part> of the universe, and such is the centre, be guarded, the Pythagoreans call the centre <and>][240] the fire which occupies this space the guardpost of Zeus.

It was a matter of concern to Plato and – in emulation of him – to Aristotle not to be thought to condemn older ideas without allowing them a defence, and so they first described earlier doctrines in a plausible way, when it was possible. Aristotle does this here when he says that many other people might also accept not assigning the central space to earth, if a person does not get his belief from the nature of things but from the persuasiveness of arguments. For it is possible to argue in a plausible way that fire, not earth, is in the centre. And he next sets out a plausible argument, first assuming as an axiom that the most honourable sublunary space is suitable for the most honourable sublunary body. And he produces a syllogism in the first figure as follows:

(i) Fire is the most honourable body;
513,1 (ii) the most honourable place is appropriate for the most honourable body;
(iii) and the conclusion is that the most honourable place is appropriate for fire.

However, the centre is the most honourable place, since a limit is most honourable; and in the cosmos the extremity and the centre are limits;[241] consequently, of sublunary places the centre is the most honourable;[242] so, if the most honourable place is appropriate for fire, but the centre is the most honourable sublunary place, then the central place is appropriate for fire.

Alexander says that Archedemus,[243] who was later than Aristotle, was of this opinion, but that it is necessary to search in history to find out who thought this before Aristotle. But perhaps <Aristotle> does not say 'many others might also accept' meaning that some others besides the Pythagoreans thought in this way, but that other people might also accept this when the argument is set out in a quite plausible way.

(293b1) And that is why he next sets out the demonstration of the Pythagoreans since they also tried to give this demonstration in a plausible way. For, he says, the Pythagoreans did not say that the earth moves around the centre and fire is situated at the centre just in order to fill out the ten bodies which move in a circle, but also they said that fire is situated in the centre 'because it is most suitable that the most authoritative <part> of the universe be guarded', and the centre, that is, what is in the centre, is such as to be guarded so that nothing foreign gets near to it and so that the central point enjoys to the greatest extent the binding and watchful uniqueness of the gods. So, because fire is guarded in the centre they call fire the guardpost of Zeus since it is guarded by the demiurgic binding power which is at the central point.

And it is also possible to call the centre itself 'the most authoritative <part> of the universe' (he previously[244] called this most honourable, as being an extremity); and it is most suitable[245] that this be guarded, and that what does the guarding be the fire which occupies this space – and they call this fire the guardpost of Zeus because it guards and not because it is guarded. So the previous argument for placing fire in the centre was based on the idea that the more honourable should be in the more honourable space, and this one on the idea that it is a guard in what most ought to be guarded. With this[246] interpretation it is necessary to take the words 'call … the guardpost of Zeus' as detached from what follows, but what is said next seems to harmonise more with the interpretation which takes the centre and not fire as what is most authoritative.[247]

293b4-16 <They argue> as if 'centre' had only one sense [and the centre of a <thing's> magnitude were the centre of the thing and of its nature. However, just as in the case of animals the centre of the animal and of its body are not the same, so one should even more make this assumption in the case of the whole heaven. For this reason they should not be alarmed about the universe or introduce a guardpost at its central point; rather they should seek *that* centre and find out what it is like and where it really is. 514,1

293b11 For that centre is a starting point and honourable, but the spatial centre resembles a termination more than a starting point, since the centre is what is delimited, the limit is what delimits. And what contains and the limit is more honourable than what is bounded. For the one is matter, the other the substance of the structure.]

Some people, then, have this opinion concerning the place of the earth ….

The first argument assumes that the centre is the most honourable space, the second that it is the most authoritative <part> of the universe, and each infers from this that fire is in the centre. He now gives a refutation of what is said on the basis of the homonymy of 'centre'; he says that the argument proceeds as if 'centre' were used in one way only and there were no difference between the centre relative to a magnitude and that relative to the nature of a thing. However, in the case of animals these two are not the same; rather in their case the centre of the corporeal magnitude is that from which the magnitude in each direction is equal (the navel, for example), and the natural centre is that in which what is most honourable and the starting point for being is (for example, the region around the heart, where this starting point or something analogous to it might be). In the same way in the case of the whole cosmos it turns out that the central point is the centre of the spherical magnitude and body, but it is necessary to seek something else as the most honourable <part> analogous to the heart, namely the centre; and this is not the central point but rather the fixed sphere because it is the starting point of the being of the cosmos and carries around the other spheres with it and contains the whole corporeal nature. Here is where one should seek what is most honourable, here where there is no need for a guardpost; and one should not say it is the central point and be alarmed about it because it needs a guardpost and therefore introduce fire at the central point, calling fire the guardpost of Zeus. So, when he said previously that 'it is most suitable that the most authoritative <part> of the universe be guarded', he meant that the centre is guarded and fire is the guard.

(293b11) And he also says that 'centre' has two senses, and the centre related to nature is really a starting point and honourable, but the spatial centre resembles a termination more than a starting point. He proves these things as follows. The spatial centre (most of all in the case of spherical figures) is delimited and contained, being bounded by the starting point, whereas the starting point delimits and contains; so if what is delimited and bounded is analogous to matter, what delimits and bounds to form and being, and form and substance are more honourable than matter, then the starting point is more honourable than the centre.

In general someone who chooses to make these hazardous comparisons should also say that demiurgic creativity proceeds from above from the fixed sphere and moves down from the heaven to the sublunary world and here introduces higher elements which are
515,1 stronger than lower ones, so that at the last it produces earth and the central point of the universe. But perhaps one should conceive of the whole heaven as a complete sphere down to its central point and so say that in the heaven the partlessness of its central point and what holds it together and what is reverenced and sovereign and its centre and its perimeter have equal standing, and that perhaps the Pythagoreans were zealous to make apparent what is divine in the central point, something which escapes the notice of the wicked.

In general <Aristotle> said[248] that the central point is more honourable because it is both a starting point and a limit, and he said that, like the heaven, it is more honourable than what is in between. And so the whole argument runs as follows:

(i) A limit is more honourable than what is between;
(ii) the extremity and the centre are limits.

Consequently the comparison is not of the central point with the heaven, but of both extremity and centre with what is between. For also Plato does not say that earth, which is in the centre, is 'the first and most senior thing' of heaven but of 'whatever has come to be inside heaven'.[249]

> **293b16-30** Similarly in the case of rest and motion, [since not everyone has the same conception. Rather those who say that it does not lie at the centre <say> that it moves in a circle around the centre, and not just it but also the counterearth, as we said previously.
>
> 293b21 And some people even think that there may be several bodies of this kind moving around the centre, but invisible to us because of the interposition of the earth;[250] and they say that is why there are more eclipses of the moon than of the sun, since each of these moving bodies, and not just the earth, blocks it.
>
> 293b25 For they think that even if we do not dwell at the central point nothing prevents the same phenomena from resulting as would if the earth were at the centre, since the earth is not the central point, but is distant from it by a whole hemisphere; for even now nothing makes it obvious] that we[251] are at a distance of half a diameter from the central point.

After his historical discussion of the position of the earth, he recounts the opinions about its motion and rest. He says that those who say that the earth does not lie at the centre, as the Pythagoreans do, say that it moves in a circle around the centre, and not only it but also the counterearth.

(293b21) And some people think that there is not just one counterearth but that 'there may be several bodies of this kind moving around the centre', but that, just as the Pythagoreans said that the counterearth is invisible to us because of the interposition of the earth, so too for the same reason these things are not seen from where we are. (He does not say who had these opinions, but Alexander says that it is possible that it is to be understood that certain Pythagoreans were of this opinion.) They produced as evidence for this doctrine the fact that, compared with the sun, the moon is eclipsed often; for it is eclipsed often because it is blocked not only by the earth but also by these other bodies which move around the centre.

516,1 (293b25) But the fact that the phenomena occur as if the earth has the relation of a central point and is at the centre told against those who said that the earth is not at the centre but outside it, whether it stands still or moves; for if the earth is not in fact at the centre of the universe, having the relation of central point to the heaven, the same stars would not seem to have the same size to observers at different places as if the observers were in the same place; and also the planes extending through our eye, which we call horizons, are observed always to bisect the heavenly sphere and greatest circles in it, since we always observe six signs of the zodiac above the earth while six remain beneath the earth, which would not happen if the earth had some perceptible size in relation to the heaven or were not in the centre. And so, as if releasing these and similar difficulties from those people, Aristotle says that the phenomena can also be preserved in terms of this kind of opinion just as they can also be according to the people who say that the earth is at the centre. For the people who say it is at the centre do not say it is the central point of the universe, since, unlike a central point, it is not without parts, since it has a certain size; rather, according to these people the central point of the universe is not the whole earth but the central point of the earth. And so the surface of the earth, from which observations are made and the planes of the horizons are extended, will be distant from the central point of the universe by a whole hemisphere of the earth. So if there is no difference as far as the phenomena are concerned for us who live on the surface of the earth and are distant by a hemisphere from the central point and <what they would be> if we were at the central point, nothing prevents the phenomena being preserved even if one were to suppose that the earth is distant from the central point by a whole sphere, as <it is for> those who say that it is outside the central point. For if the earth were in the centre and we lived at the central point, there would be a great difference relative to those who say that the earth is outside the centre, but if the earth is hypothesised to be in the centre but we do not live at the central point but on the surface, the surface of the earth will not differ much whether the earth lies in the centre or outside it. For whether the earth lies outside the centre or in it, the surfaces which face towards each other can be in practically the same place.

517,1 **293b30-32** And some people say that it lies at the central point [and is wound and moves around the pole which stretches through the universe,] as is written in the *Timaeus*.[252]

Having spoken about those who say that the earth lies outside the central point and moves around it, <Aristotle> adds that some assert that it lies at the central point and winds and moves, as is written in the *Timaeus* of Plato. The words of Plato's *Timaeus* are the following:

> <The demiurge> designed earth, which is our nurse,[253] wound[254] around the pole which stretches[255] through the universe, to be the guardian and demiurge of night and day, the first and most senior of the gods which have come to be inside heaven.[256]

He is now calling the axis the pole[257] which stretches through the universe. In Plato the word 'pole' means three things: in the *Phaedo* he calls both the heaven and the limits of the axis around which the heaven turns, poles,[258] and here in the *Timaeus* he also calls the axis a pole. The word *illomenên* (wound), if it is written with an 'i', indicates that the earth is bound, and this is the way Apollonius the poet uses it:

> <Hercules> shook the beast, wound in bonds, from his great back.[259]

And so does Homer:

> They wind it in bonds[260] and drag it unwilling by force.

But if it is written with the diphthong 'ei' (*eillomenên*) it then indicates being shut in, as it does with Aeschylus in the *Bassarids*.[261] And that Plato has used *illomenên* is made clear by what is said in the *Phaedo* about the earth:

> For a thing evenly balanced and placed in the centre of something uniform will not be able to incline more or less in any direction.[262]

It is also made clear by what Timaeus <Locrus> (whom Plato is here following) says:[263]

> Earth, seated in the centre, is the hearth of the gods, boundary[264] of night and dawn,[265] producing settings and risings[266] at sections of the horizons.[267] It is the most senior of the bodies within heaven.

And then, a little later, in speaking about the other elements, he adds:

> So the earth is the root of everything and base for the others and is itself fixed firm by the same inclinations <towards the centre>.

> Alexander says: 518,1

> Since Aristotle asserts that this is what is said in the *Timaeus*

and the text itself, *illomenên*, indicates turning,[268] to declare by force that it is not expressed[269] in this way is a case of changing what is said in the direction of one's own hypotheses. For even the word 'rolled up'[270] into which they change 'wound' is itself indicative of moving. And if elsewhere (and here he is obviously pointing to what is in the *Phaedo*) Plato speaks in a different way, this is irrelevant to the argument. For Aristotle is chastising what is in the *Timaeus*, whether Plato is there expressing his own view or the view of Timaeus <Locrus>.

Alexander says these things, and I will start with the last texts. That Timaeus <Locrus> says that the earth is seated in the centre and is a basis and is fixed firm is prima facie clear from his words which I have set out. The text makes clear enough that the earth is bound – as the words[271] set out indicated, as does the word 'rolled up' – when the text is written with the diphthong *ei* and is written with one *l*;[272] and, furthermore, being rolled up is also applied to circular figures even if they are motionless. In addition, even if the text might indicate either being bound or moving, how ought one to understand the words of Plato when he adheres to the statements of Timaeus that things are this way, and he has demonstrated[273] the stationariness of the earth: as indicating that the earth is bound or that it is in motion?

But <Alexander> says:

If Aristotle asserts that <Plato> has spoken in this way, it is not reasonable to speak against him, since it is not really likely that he was ignorant of either the meaning of the text or of Plato's intention.

So perhaps, all things considered, it would be correct to say one of two things:

(1) Since the text <can> also indicate being rolled up, Aristotle takes the text in this way, just as he usually concerns himself with refuting the surface meaning of texts because of those who understand things in a more superficial way. For, then,[274] if he were speaking <about Plato> after speaking about those who say that the earth[275] moves around the centre and someone, understanding *illomenên* as indicating movement, were to suppose that Plato says the earth moves at the centre, there would be at the same time another hypothesis made by people who say that the earth moves (for it would be hypothesised to move either around the centre or at the centre), and the person who understands what Plato says in this way will be refuted at the same time.[276]

(2) Or one should say that the words 'and moves' were added

to 'is wound' <at 293b31> later by someone,[277] but that Aristotle having originally proposed to inquire about the earth, whether it is stationary or moving,[278] and having given a historical 519,1
account of those who say it moves around the centre, has added the witness of Plato's *Timaeus* that it is bound and stationary at the centre; for if 'and moves' is not added, it is also possible to understand what he says in this way. That he has given a historical account in terms of oppositions, not just in the case of moving but also in the case of being at rest, is also made clear by what he adds about the shape <of the earth>: that people speak in much the same way as in the case of motion and rest, some saying that it is spherical in shape, others that it is drum-shaped.[279]

Heraclides of Pontus thought he could preserve the phenomena by hypothesising that the heaven is stationary and that the earth is at the central point and moves in a circle.[280]

293b32-294a11[281] They dispute in much the same way about the shape <of the earth>, [since some people think that it is spherical and others that it is flat and drum-shaped. They give as evidence for <the latter view> that the sun, in setting and rising, is observed to make the part hidden from sight by the earth a straight line rather than a curved one on the grounds that if the earth were spherical it would be necessary that the section be curved. But these people do not take into account the distance of the sun from the earth and the size of the circumference; for from far away it appears to be a straight line in the apparently small circles <of the sun's disc>.[282] So they should not at all doubt that the bulk of the earth is round on account of this appearance.

294a9 But they add more and say it is necessary for it to have this shape in order to be stationary.

294a10 And so the ways in which people have spoken about the motion and rest of the earth are many.]

And he says that there is a different history to tell about the shape, since some people say the earth is spherical, others that it is flat and drum-shaped. He says that those who say it is drum-shaped give as evidence for this that the sun, in rising and setting, is observed to have a straight line distinguishing its visible and invisible parts. However, they say, if the earth were spherical and stood in front of the sun, the section would be crescent-shaped or doubly convex, as is seen in the case of solar and lunar eclipses: in the case of solar eclipses the moon blocks <the sun>, and in the case of lunar eclipses, the cone of the shadow <of the earth is responsible>. But they say that since we are, in fact, on the plane <surface> of a drum-shaped figure we see the sections as straight lines. For if the arc of a circle is placed in the same plane as the eye, it will appear to be a straight line, as has been proved in the *Optics*.[283] Against the people who assert these things Aristotle says that they do not take into account the great distance of the sun from the earth and the very small apparent size of the solar body in relation to its distance or that the circles in apparently small bodies appear to be straight lines from a great distance. For spherical surfaces are judged to be plane from far away, as in the case of the sun and the moon.

But if the distance and apparent smallness of size is the cause of the section appearing to be a straight line, why doesn't the same thing happen in the case of solar and lunar eclipses, since the distance and the apparent size are the same? Perhaps one should say that if we 520,1 were outside the earth and saw the sun partially obstructed by the earth, the sections would always appear to us to be curved, just as do the sections of the sun in solar eclipses, when it is obscured by the moon, or of the moon when it is obscured by the cone of the shadow <of the earth>. But now, since we are on the earth and we see the sun rising and setting because of the horizon (and a horizon is the plane extended through the surface of the earth and our eye), it results that the solar sphere is cut in a circle by the horizon; but a circle which is in the same plane as our eye is seen as a straight line.

(294a9) Having rejected the evidence based on the sun as an insufficient ground for supposing that the earth is not round, Aristotle says that they add another reason why the earth is drum-shaped when they say that it is necessary for it to have this shape in order to be stationary, since it is stationary because it has this shape. For a spherical shape moves easily, but the shape of a drum is suitable for rest since it is mounted on what is under it with an entire plane <surface> or, as they say, it sits like a lid on the air beneath it.[284]

(294a10) Having said that the earth is stable because of its drum shape, he adds that the ways in which people have spoken about the arrangement[285] of the earth are many. On some views its arrangement is suitable for motion, on others, such as the one just mentioned, it is suitable for rest.

294a11-21 That there is difficulty must strike everyone. [For a mind must be quite untroubled not to be surprised at how a little piece of earth which is raised up high and released moves and is unwilling to remain at rest, and a larger piece always moves faster, but if someone were to raise up the whole earth and release it, it would not move. But in fact it is stationary, even though it is so heavy. However, even if someone were to take away the earth when parts of it were moving but before they had fallen, they would move down if there were no resistance.

294a19 As a result the difficulty has reasonably enough become a subject of philosophical discussion for everyone. But one might be surprised that the solutions concerning this do not seem <to people> stranger than the difficulty.[286]]

He has already recounted the opinions about the position, motion, and shape of the earth, and briefly spoken against them, disdaining most of them as inconsistent; he next transfers the discussion to those who say that the earth is at rest but do not provide a good explanation of its being so. And first he mentions those who say it is at rest because the earth is infinite,[287] such as Xenophanes of Kolophon; second those who say it rests by being supported on water, as Thales said; third those who say it rests by being held up by an underlying air on which the earth, being flat and drum-shaped, sits like a lid and does not agree to withdraw – Anaximenes, Anaxagoras, and Democritus are thought to have spoken in this fashion;[288] fourth the associates of Empedocles, who make the vortex of heaven responsible for the earth being at rest; and fifth those, such as Anaximander and Plato, who say that the cause of its being at rest is its uniformity and even balance.

But first he constructs the difficulty relating to the earth's being 521,1
at rest, that it must strike everyone that there is a difficulty about how the earth remains at rest. 'For a mind must be quite untroubled', that is, lazy, 'not to be surprised at how a little piece of earth which is raised up high and released moves' down 'and is unwilling to remain at rest' and a greater piece always moves faster, but if someone were to raise up the whole earth and release it, it would not move down. Satisfactory evidence that it would not move if it were released is the fact that now it is stationary although it is on high and has such great weight. And he says that this is a difficulty worth raising concerning the same thing: how if the parts of the earth were falling from above, if someone were continually to move away the underlying earth before they had fallen, they would remain in motion until something resisted them, although the whole earth does not move but is stationary even though there is nothing resisting beneath it.

(294a19) He says that because of these things the difficulty about the earth has become a subject of philosophical discussion for everyone, since it is genuinely philosophical to seek the causes of existing things. But concerning the solutions of these[289] one might be surprised <that they> continue[290] to seem stranger than the difficulty, since, although the <people who offer them> wish to solve the difficulty in these matters, they seem to use things which are worthy of more difficulty. Alexander offers this interpretation, writing the text as 'but concerning the solutions of these[291] one might be surprised' <and so forth>; he explains that the text lacks completeness, as has been said.[292] But perhaps, even if this is the text, its expression is brief rather than elliptical.[293] For, having said that 'the difficulty has reasonably enough become a subject of philosophical discussion for everyone', he adds, 'but concerning the solutions of these <people>[294] one might be surprised that they (obviously the people who offer the solutions) do not seem stranger than the difficulty'. So one should not attach the word 'strange' to the solutions, but to the people who offer them. However, I have found that many of the copies read this way: 'but one might be surprised that the solutions concerning these <things> do not seem stranger than the difficulty.' And this is smoother.

Having said these things he next says which of the earlier people provided which explanations for the earth remaining at rest.

522,1 **294a21-b13** For these reasons some people, such as Xenophanes of Kolophon, say that what is beneath the earth is infinite,[295] [so that they don't have the concern of seeking the explanation. And so Empedocles chastises them, saying:[296] 'if the depths of earth and abundant *aithêr* were infinite, as has been said by the tongue of many people, and poured out from their mouths, people who have seen little of the universe'.[297]

294a28 Others say it lies on water. This is the earliest theory which has come down to us; they say that Thales asserted it on the grounds that earth remains at rest because it floats like a piece of wood or something else of that sort (since none of these things is of such a nature as to remain on air, but <only> on water). As if there was not the same issue concerning earth and concerning the water supporting the earth; for water is not of such a nature as to remain on high, but it rests on something.[298]

294b1 Furthermore, just as air is lighter than water, water is lighter than earth; so how can what is lighter lie beneath what is heavier by nature?

294b3 Furthermore, if the whole <earth> is of such a nature as to remain on water, it is clear that each of its parts will be so too; but, in fact, this is not observed to be the case, but a chance piece moves to the bottom[299] and more quickly the greater it is.

294b6 These people seem to have inquired up to a certain point but not as far as the difficulty allows. And it is habitual for all of us not to inquire with respect to the issue, but with respect to the person who says the opposite. A person inquires within himself up to the point where he is no longer able to argue against himself. So the person who is going to inquire correctly should be able to raise objections based on what is appropriate to the <subject> genus, and this comes from] having studied all its differentiae.

He says that all these people are moved by the difficulty but are not willing to have the concern of seeking until they find the most authoritative explanation, and each of them declared what struck them as obvious, just as Xenophanes of Kolophon declared that what is beneath the earth is infinite and as a result it remains at rest. Since I have not encountered words of Xenophanes dealing with this subject, I do not know whether he said that the underneath part of the earth is infinite and as a result it remains at rest or that the region under the earth and *aithêr* is infinite so that the earth falls *ad infinitum* and seems to be stationary. For Aristotle does not make this clear and the words of Empedocles do not determine this clearly, since 'depths of earth' might also mean the things into which it descends.

(294a28) However, he does not consider this opinion worth arguing against since it is completely implausible, and after it he sets out the opinion of Thales of Miletus, who says that the earth is supported on water like a piece of wood or something else of such a nature as to float on water. Aristotle argues against this position, which is perhaps more prominent because it is expressed by the Egyptians in the form of a myth. And perhaps Thales took the doctrine from there.

The refutation consists of three arguments. One derides them all elegantly for trying to solve the difficulty by means of something no less difficult, since someone could raise the same question about the water underlying the earth: what does it stand on when it supports the earth? 'For water is not of such a nature as to remain on high', just as earth isn't either.

(294b1) The second argument is the following. If lighter things are of such a nature as to rise up over heavier ones, and heavier things support lighter ones, and, just as air is lighter than water, water is lighter than earth, 'how can what is lighter lie beneath what is heavier by nature'?

(294b3) The third argument uses what is called a conversion with antithesis.[300] If the whole earth is of such a nature as to remain at rest on water, it is clear that each of its parts can ride on water. However, this is not observed to be the case, since a chance piece which is thrown into water does not stand still until it reaches the depth of the water (this is what 'sinks' means), and a greater piece moves more quickly. From this it is clear that if you put water under
earth, the whole earth would sink down faster than any part of it. The 523,1
upshot of the syllogism is obvious: therefore the whole earth is not of such a nature as to remain at rest on water.

(294b6) Having said these things, he discloses in a wonderful and universal way the reason we do not solve difficulties well. For, he says, 'these people seem to have inquired up to a certain point' but not to the extent of the difficulty, but only to the point where it would still be possible to raise difficulties for such solutions; so they are satisfied with giving a random solution. He confirms that this is the way things are by saying that it is habitual for all of us not to inquire with respect to the issue set out, 'but with respect to the person who says the opposite': for if we satisfy this person, we are contented even if the person does not put forward to us all the difficulties which are appropriate to the issue. And again he confirms that this is the way things are from the things which each of us is conscious of undergoing by himself in inquiries; for each of us by himself inquires about an issue set out up to the point where he is no longer able to argue against himself, since then we think that no difficulty remains. As a result the person who is going to inquire correctly should consider all the difficulties which are appropriate to the issue set forth and raise objections against all of them, whether he makes his inquiry to himself or to another person, since in this way the solutions of the difficulties will have completeness.

He calls objections appropriate if they are taken from the issue itself and are not sophistical. We should raise difficulties and objections against all the objections which are appropriate to the issue if we are to have considered all the differences by which the issue set forth for inquiry differs from the others. For the people who say that the earth floats on water because they saw wood floating on it were deceived because they could not make out the difference between earth and wood. He also says in the first book of the *Topics*[301] that one of the things which is serviceable for furnishing arguments for the sake of understanding is reflecting upon all the differences from one another of those things which most seem to have community with one another.

In saying this now he has passed over acknowledging that one should also study all of the points of communion of things that differ. For many difficulties are set in motion by this. But the things which are said in the *Topics*[302] to complete the discovery of premisses for the furnishing of syllogisms are: making divisions for what has several senses, finding the differences appropriate to the issue set forth, and the study of similarity.

524,1 **294b13-23** Anaximenes, Anaxagoras, and Democritus [say that the cause of its remaining at rest is flatness. For it does not cut the air beneath it but sits on it like a lid, as those bodies which are flat are observed to do, since they also are hard to move against winds because of the resistance. They say that because of its flatness the earth behaves in the same way with respect to the underlying air which, not having sufficient room to change position, is stationary because of a mass <of air> underneath it[303] – this can be compared with the water in clepsydras.[304] And they give a lot of evidence that air which is cut off and remains fixed is able to support a great weight.]

Third he examines those who say that the cause of the earth's remaining at rest is its flatness. For this flatness, even if it is of such a nature as to move downward, is not able to cut the underlying air, but it sits on it like a lid, as bodies which are flat are observed to do; for they are seen to sit like a lid. For, just as bodies which are flat are hard to move against winds because they block and resist winds (for when the wind is blocked and has no way out it remains motionless, as in the case of houses with no ventilation, and as a result flatter bodies are hard for the wind to move), so the earth because of its flatness behaves in the same way with respect to the underlying air, but the air, not having sufficient room for changing position is constrained to be stationary by a mass underneath which is part of it and does not move; for since the part next to the earth does not have a way out, what is under it remains as a mass, and so all the air is stationary with a mass <of air> underneath, and if it is stationary and is not divided, it is necessary that the earth remain at rest on top of it.

He introduces the water in clepsydras as an example of the air beneath remaining motionless when the air near the earth does not have a passageway. A clepsydra is a container with a narrow mouth and a wider bottom pierced by little holes; it is now called a *hudrarpax*.[305] When the container is lowered into water with its mouth above stopped up, the water will not enter through the holes because the mass of the air in the container resists the water and prevents its upward passage since it does not have any place where it can change position. But when what is stopping up the mouth above is taken away, the water will enter as the air makes way for it. Thus it is in this respect[306] that the example provides a resemblance: when the water is taken as analogous to the earth which does not move through the underlying air and what is in the clepsydra is taken as analogous to the air which underlies the earth and resists it. But in addition if the clepsydra is filled with water and one stops up the mouth above so that air does not enter,[307] the water does not flow out through the holes in order that the container remain not void when air does not enter through either the stopped up mouth above or through the holes below which are the way out for the water through their entire opening. So, just as, when the air does not enter, the water in the 525,1
clepsydra remains at rest, so too the air under the earth remains at rest because it does not have a passageway; but if it remains at rest the air under it also remains as a mass, and when all of this remains at rest the result is that the earth is supported and held up by it.

In connection with Aristotle saying that they give a lot of evidence that the air which is cut off and remains fixed is able to support a great weight but not himself introducing the evidence, Alexander says that it is clear that the air is strong because of earthquakes and winds and the fact that the strongest things are torn asunder when the sound from thunder crashes into them.[308] But it is clear that these things are acts of a blast of air which is moving and not stationary. So the example based on skin bags which he sets out in the sequel[309] is better, since when these are blown up and supported by water they carry heavy burdens, as I also have seen in the case of the river Aboras.[310]

294b23-295a29 First of all, if the shape of the earth[311] is not flat, [it cannot be stationary because of its flatness.

294b24 However, on the basis of what they say, it is not its flatness but rather its size which is the cause of its being at rest. For the air, not having a passageway because of its confinement remains fixed because of its quantity: it is large because it is enclosed by the large magnitude of the earth. So, this would be the case even if the earth were spherical and just as large in size, since it would remain fixed according to their argument.

294b30 In general our dispute with those who speak in this way about the motion of the earth is not about parts but about a certain whole and entirety. For one should determine at the start whether there is some[312] natural motion in bodies or none and whether there is not a natural motion, but there is a forced one. But since we have determined this previously[313] to the extent of our present abilities, we must use these determinations as facts. For, if they have no natural motion, they won't have a forced one either, but if there is neither natural nor forced motion, nothing will move at all. For concerning these things it has been determined previously that this must result, and it has also been determined that there cannot be rest either, since just as motion is either forced or natural, so too is rest. However, if there is some natural motion, there will not be either forced motion only or forced rest only. So if the earth now remains at rest by force, it also came together at the centre because of the vortex, since, on the basis of what happens in the case of things in liquids or air, everyone says that the vortex is the cause; for in liquids or air greater and heavier things are always carried towards the centre of the vortex.

295a13 And so everyone who says that the heaven came to be says the earth came together at the centre,[314] and they seek the cause of its remaining there. And in this way some of them say that the cause is its flatness and size, others, such as Empedocles, that it is the motion of the heaven which, running around in a circle and moving faster than the earth, prevents <the earth from moving>, like water in ladles; for, when a ladle is moved in a circle, although it is of a nature to move downward and it is often underneath the bronze, it does not move downward for the same reason.

> 295a21 However, if neither the vortex nor flatness prevents <the earth from moving>, but instead the air yields,[315] where will the earth move? For it moves to the centre by force, and it remains there by force, but it is necessary that it have some natural motion. Is this motion upward or downward or in what direction? For it is necessary that it have some natural motion, but if that motion is no more upward than downward, the air above will not prevent its motion upward, nor would the air under the earth prevent its motion downward;] for it is necessary that the same things be the cause of the same things for the same things.

Having presented in as plausible a way as possible the case of those who say that the earth is stationary because of its flatness, he now argues against them; and he first uses the following argument:

> If someone says that the earth is stationary because of its flatness and a shape of this sort, then, if it were proved that it is not flat but spherical, flatness could not be the cause of its being stationary. And so it is necessary that the person who argues in this way prove first that the earth is flat and, in this way, that it is stationary because of its flatness.

Perhaps Aristotle has converted the syllogism which they state, which is the following:

> If the earth remains at rest, it is flat, and it remains at rest because of its flatness.

And he says:

> If the earth is not flat, it could not be stationary because of its flatness.

The result is that, because it has been proved that the earth is not flat, it is also demonstrated that it is not stationary because of its flatness.

(294b24) Secondly, it is evident on the basis of what they say that even if the earth remains stationary because it is held up by the underlying air, its flatness is not the cause of its being at rest; rather its size is. For if the <amount of> air is large and fills the whole space and it is confined because it is contained by the large magnitude of the earth, it would not move in this space. But if it also has no passageway because it is sat on as if by a lid at every point, it would 526,1
not exit either. But it is possible that these things be the case even if the earth is not flat but spherical and so large in size as to sit like a

lid on the air at every point; for in this way the air would also remain motionless with the earth on top of it, according to their argument. So, even if the earth is in fact spherical, they are able to provide this explanation because of its size, but they limit it to the flatness of its shape. So neither those people <who say the earth is flat> nor the person who proves that the earth is spherical has correctly stated[316] the cause of its being at rest on the basis of the underlying air.

(294b30) Having overturned their argument in this fashion, he sets out a third argument which starts in a universal way from principles of physics. He says that in general his dispute with those who speak in this way about the motion of the earth and say it does not move is not about the earth itself, which is a part of the universe, but about the whole structure of natural bodies; for it is necessary that what holds of other things also hold of the earth. And so one should determine at the start whether there is a natural motion in bodies or there is none, and, if there is not a natural motion, then whether there is a forced motion or there is not a forced motion either. And, having made this division, he reminds us of what was said in the *Physics* and in the preceding book,[317] where it has been proved that if there is no natural motion there will not be a forced one either. For an unnatural motion is forced, and the opposite of a natural motion is an unnatural one; so if the unnatural is a turning away from the natural, if there were no natural motion there would not be unnatural or forced motion; so if there were no natural motion, nothing would move at all. And the same things have been proved in the case of rest: if there is unnatural rest, it is necessary that there also be natural rest, and if there is no natural rest, there isn't unnatural rest either, nor is any body at all stationary. But it was also proved that if something doesn't move naturally it isn't stationary either naturally or unnaturally. For natural rest is the limit of natural motion, as unnatural rest is of unnatural motion; consequently, if there is no natural motion nothing either moves or is stationary. If these things are absurd (since many things are observed to be moving and stationary), it is clear that bodies have a natural motion. But if they do, 'there will not be either forced motion only or forced rest only'. So if, according to those who believe that the earth is supported by the air, the earth now remains at rest at the centre by force, it is clear that it is carried to the centre by force, as those who compress it forcibly into the centre by means of a vortex say.

(295a13) He next inserts the statement that everyone who asserts the coming to be of the cosmos says that the earth came together in the centre because of the vortex motion of the heaven. And this is not
527,1 just Empedocles, but also the associates of Anaxagoras and others. And they are carried[318] to this position because of the fact that rotations in water and in air forcibly compress greater and heavier bodies towards the centre, and they make the vortex responsible for

the initial motion. In seeking[319] the cause of its remaining there, some say its flatness is the cause, as was said previously,[320] others, such as Empedocles, say that the vortex motion of the heaven[321] is also the cause of this, since the motion of the heaven, running in a circle and moving faster than the downward motion of the earth, prevents it <from moving>, 'like water in ladles'. For when ladles are made to revolve with extreme speed by jugglers (and this is now done with cups), although it is of a nature to move downward and the water is often beneath the bronze of the ladles or cups, it does not move downward because of the vortex motion; rather, because the vortex motion is faster than the natural inclination of the water, it prevents it from moving.

(295a21) Having inserted these statements about the earth coming together at the centre by force and remaining at the centre by force according to those who make the vortex and the earth's flatness responsible for these things, he next adds the consequences of what was said before. If the earth came together to the centre by force and remained there by force, if the force were hypothetically taken away so that neither the vortex nor its flatness prevented it <from moving>, but also the supporting air were to depart,[322] where would the earth move naturally? For they say that it came together at the centre by force and it remains there by force, but because of what has been proved, it is also necessary for it to have some natural motion. So is this motion upward or downward or in what direction? For it is necessary that it have some natural motion. And if it was carried to the centre by force from every direction, it will also naturally move in every direction from the centre. But if that motion is no more upward than downward (because it is in every direction), the air will not prevent its motion upward; for they do not make the air lying above responsible for its not moving upward, so that the air should not be the cause of its not moving downward either. 'For it is necessary that the same things be the cause of the same things for the same things.' So if the earth naturally moves upward and downward in the same way (which is what follows for those who say that it is carried to the centre by force and remains in the centre by force), whatever is the cause of its not moving upward naturally should also be the cause of its not moving downward naturally either.

Having said that both the associates of Anaxagoras and Empedocles say that the same thing, the vortex, is the cause of the earth's coming together in the centre, but that the former make its flatness responsible for its rest there, the latter makes the vortex responsible, and having spoken against those who make the vortex and flatness and the resisting air responsible for its rest, he turns the discussion back to Empedocles.

528,1 **295a29-32** Furthermore, one might say this against Empedocles. [When the elements had been split apart by Strife, what was the cause of the earth's rest?] For he cannot also make the vortex responsible at that time.

Having just spoken together both against those who make the vortex responsible for the earth's rest and against those who make the earth's flatness responsible, he adds another argument against Empedocles, saying that 'furthermore one might say[323] this against Empedocles'. He uses four arguments, the first of which seems to have been expressed unclearly. He says:

> When the elements had been split apart by Strife, what was the cause of the earth's rest? For he cannot also make the vortex responsible at that time.

Now with the words 'when the elements had been split by Strife' he seems to refer to some condition besides the present one and calling that the one which comes to be because of Strife. However, Empedocles says that this cosmos comes to be because Strife divides the elements, just as the Sphere comes to be when Love brings them together and unites them. So how can <Aristotle> say that there is no vortex under the domination of Strife, since that domination is <now>?[324]

Alexander thinks that he can somehow straighten out what is said in the following way:

> When the elements had been split apart by Strife but had not yet come to be separate – that is, had not as yet been separated and divided into this cosmic order in which the vortex is the cause of the earth's resting in the centre – but rather when they were together with Love dominating, what was the cause of the earth's resting then? For there was no vortex at that time, since then the elements stood still, not yet having been divided apart by Strife in this way. So, one should speak in this way or[325] say that 'split apart by Strife' means 'Strife having been separated from them'. For Strife is assumed by Empedocles to be the cause of a cosmic order of the elements like the present one and of the vortex motion of the surrounding air, which, he says, moves in a circle when it has first been separated out.

That <Alexander> has understood and combined these words in an implausible way is manifest to anyone, since he understands Aristotle to be speaking of the elements as not having come to be separate, and he forcibly tries to make the time referred to by the clear words of Aristotle 'when the elements had been split apart by Strife' be the domination of Love. And who would use the words 'when the elements had been split apart by Strife' to mean 'Strife having been separated from them'? However he was forced or rather is forced[326] because he believes that according to Empedocles this cosmos has come to be because of Strife only. But perhaps, even if Strife dominates in this cosmos, as Love does in the Sphere, nevertheless both <this cosmos and the Sphere> are said to come to be by the action of both <Love and Strife>. But perhaps there is nothing to prevent setting out some of the words of Empedocles which make this clear:[327]

> But I shall turn back to the path of song I traced before, drawing 529,1
> off this argument from another:[328]
>
> When Strife had reached the lowest depth of the vortex and Love had come to be in the centre of the eddy, in Her then all these things come together to be one thing only; not immediately, but different things coming together from different directions at will. And, as they were being mixed, countless types of mortal things poured forth, but many things that Strife still restrained from above stayed unmixed, alternating with those which were combining, for It had not yet[329] blamelessly and completely retired to the last limit of the circle, but It remained in some of the limbs and departed from others. And to the degree that It was always running out, a gentle, immortal impulse of blameless Love followed in pursuit. Immediately, exchanging their paths, what had learned previously to be immortal grew as mortal, and what had previously been unmixed became mixed.

In these lines it is made clear that in the simple cosmic order Strife is confined and Love dominates when it has come into the centre of the eddy, i.e., the vortex, so that there is also a vortex when Love dominates; and it is made clear that some of the elements remain unmixed under Strife, but the mixed ones make mortal animals and plants (mortal because things which are mixed are again dissolved).

But also in speaking about the coming to be of corporeal eyes <Empedocles> adds:

> From which the goddess Aphrodite fashioned tireless eyes,[330]

and a little later:

> Aphrodite, putting <them> together with bolts of love,[331]

and, giving the reason why some things see better in the day, others at night, he says:

When they first grew together in the hands of Kupris.[332]

And <to see> that he is speaking about these things in this cosmos, hear these words:

530,1 But if your belief about these things is in any way defective: how the forms and colours of mortal things, as many as[333] have now come to be, came to be when water, earth, *aithêr*, and sun were mixed, harmonised together by Aphrodite,[334]

and a little later:

As then when Kupris, busily making forms,[335] moistened earth in water and gave it to nimble fire to strengthen,[336]

and again:

Of things which She fashioned dense inside and rare outside, they having chanced on this kind of wateriness in the hands of Kupris.[337]

I have collected and set out these passages from a few of the verses which present themselves to be taken literally. But perhaps Empedocles, speaking as a poet in a more mythical way, does say their domination is by turns:

At one time all things coming together into one thing because of Love, at another again being carried apart from each other because of the hostility of Strife.[338]

And Aristotle, contenting himself with this more mythical statement, asks those who make the vortex responsible for the earth's being at rest the following question:

Since, when the elements had been split apart by Strife, the elements were unmixed and there was no order between heaven and earth – or rather, since the elements were unmixed there was not yet, on this hypothesis, even a heaven, but there was earth, since the elements are eternal, as they suppose – what then was the cause of the earth's rest? 'For he cannot also make the vortex responsible at that time.'

I think that Empedocles would say that there is not a time when the elements had been split apart and their coordination with one another did not exist, since, if there was, they would not be elements; rather Empedocles' words are intended to unfold the nature of things, and they hypothesise the coming to be of what does not come to be and the division of unified things and the unification of divided things.

> **295a32-b10** It is also strange that they do not reflect [on the following question: previously the parts of the earth were carried to the centre by the vortex, but what is the reason why all things which have weight now move to <the earth>, since the vortex is not near us?
>
> 295b1 And furthermore, what is the reason why fire moves upward? It is not because of the vortex. But if fire is of such a nature as to move somewhere, it is clear that one should suppose that earth is as well.
>
> 295b3 But heavy and light are not determined by the vortex either; rather there are previously existing heavy and light things and the former come to the centre, and the latter rise up because of the motion. So these things were heavy or light before the vortex came to be, but by what were they determined, and how were they of such a nature as to be carried, and where? For if <the universe> is infinite, it is impossible for there to be an up or down, but heavy and light are determined by these two.]
>
> So most people work with these explanations.

Next he adds a second argument based on what they say. Those who generate the cosmos from a starting point by the action of a vortex thrust the earth away into the centre as if it were chips in swirling waters; they say that it then remains at rest in the centre by force. It is then strange, he says, that they do not reflect upon the question what the reason is why things having weight now move towards the earth. You cannot say that it is because of the vortex, since even if it exists now, the vortex is not near us. But for whatever reason heavy things now move to the centre, it is for the same reason that the earth was then carried, since that was its starting point of motion.

(295b1) He adds this third argument. If fire is not carried upward by the vortex, but because it has a particular sort of nature – for, as they say, the vortex is the reason why heavy things move towards the centre – why do we not say that the earth moves to the centre because it has a particular nature? Unless someone were to say that, although heavy things are carried to the centre by the vortex, things lighter than them rise up by necessity, and this is the reason why fire moves upward. For also Aristotle himself will say, 'the former come to the centre, and the latter rise up because of the motion'.

(295b3) He adds a fourth argument in which he draws out the absurdity on the basis of which they speak. For if they say that the vortex forcibly compresses heavy things into the centre, one should note that the vortex is not the cause of the heaviness or lightness of bodies according to them, but that heavy and light things exist prior <to the vortex> because they differ in their inclinations and therefore in the vortex one was carried in one direction, the other in another. So if some things were heavy and others light in their own nature before the vortex, how were they distinguished from one another? And what was the place proper to each, and where did they move naturally? For this is not the same, since if what it is to be heavy and what it is to be light were the same, the vortex would not separate these identical things, but they would also be carried in the same direction by the vortex. But if someone were to assign to them proper natural places, he would have the explanation for their natural motion and rest. But not all <of these people can> produce this. For if the universe is infinite, as Anaximander and Anaximenes are thought to say,[339] it would be impossible for there to be up and down, and if these did not exist, neither would heavy and light, since what moves downward naturally is heavy, what moves upward naturally is light.

295b11-296a23 There are some people, such as Anaximander among the earlier thinkers, who say it remains at rest because of uniformity, [since it is no more suitable for what is situated at the centre and uniformly related to the extremities to move upward than downward or to the side; but it is impossible to move in opposite directions simultaneously, so that necessarily it remains at rest.

295b16 This is said cleverly but not truly, since according to this argument it would be necessary that anything whatever which is placed at the centre remain at rest, so that even fire would be stationary there – for what was said does not apply uniquely to the earth. But this is not necessary, since <earth> is not only observed to remain at the centre, but also to move towards the centre. For it is necessary that the whole <earth> move wherever any part of it moves, but it also remains naturally wherever it moves naturally. Therefore, <it does not remain at rest> because of being uniformly related to the extremities, since this is common to all things, but moving to the centre is unique to earth.

295b25 It is also strange to inquire why the earth remains at the centre, but not to inquire why fire <remains> at the extremity. For if the extremity is its place by nature, it is clear that it is necessary that there also be some place for earth by nature. But if its present position is not its by <nature> but it remains there because of the necessity of uniformity, then they should inquire about the resting of fire at the extremities. (This uniformity is like the assertion that a hair which is strongly and uniformly stretched in every direction will not break or that if a person is extremely and equally hungry and thirsty and at an equal distance from food and drink, he also necessarily remains stationary.)

296a1 It is also surprising that they inquire about things being at rest but not about their motion. <They do not ask> for what reason one thing moves upward, another downward, if nothing hinders them.

296a3 However, what they say is not even true. It is indeed true in the indirect sense that it is necessary that anything for which it is no more suitable to move here than there remain at the centre.

296a6 But on the basis of this theory it will not remain but it will move, not indeed as a whole, but having been torn apart. For the same argument will also apply to fire, since it would be necessary that it remain at rest in the same way as earth if it is placed <at the centre> because it would have the same relation to any point at the extremities. However, if nothing prevented it, it would move from the centre towards the extremity (as it is also observed to do), except that <it would not move> as a whole towards one point (since this follows necessarily only on the basis of the argument about uniformity), but a proportional part would go to a proportional part of the extremity. (I mean, for example, that a fourth part would go to a fourth part of the surrounding <extremity>; for a point is not a body.)

296a17 And just as what is compressed can come together from a large place into a smaller one, so what becomes more rare can come from a smaller into a larger one. Consequently, on the basis of the theory of uniformity, the earth would move in this way from the centre if it were not the place of the earth by nature.

These, then, are practically all the ways of conceiving the shape of the earth and its position, rest, and motion.]

Plato is also of this opinion when he says in the *Phaedo*:

> For a thing which is evenly balanced and placed in the centre of something uniform will not be able to incline more or less in any direction.[340] 532,1

But Aristotle, finding that it was assumed earlier by Anaximander, thinks it more suitable to refute him than to argue against Plato; or, as some of those who set the <two> philosophers in opposition say, since he earlier said that Plato says that the earth 'winds and moves about the pole which stretches through the universe'[341] he does not refer this opinion about the earth's rest to Plato. Alexander says the following:

> By saying 'such as Anaximander among the earlier thinkers' he makes clear that there were others of this opinion who were not 'earlier'. Or rather it is unclear, as far as this text goes, whether it is stationary at the centre or whether it remains at the centre because of its own even balance and the uniformity of what contains it, but, while remaining there, it winds and moves 'around the pole which stretches through the universe'.

Anaximander thought that the earth remains at rest because of the air holding it up[342] and because of even balance and uniformity. Having proposed to speak against this opinion, <Aristotle> first sets out the argument supporting it, as is his custom:

> It is no more suitable for what is situated at the centre and uniformly related to the extremities to move upward than downward or to the side because of its being everywhere evenly balanced and because of the uniformity of it and what contains it. But it is also impossible to move in opposite directions simultaneously. So if it is not possible for this <earth> to move all together or by turns, it is necessary that it be stationary.

(295b16) He says that this argument is spoken cleverly, i.e., in a plausible way, but not truly. For being related in a uniform way to the extremities is not unique to the earth; it also attaches to fire and the other elements, since each of these, because of its own homoiomerousness and that of what contains it, is also related in a uniform way to what contains it. Consequently, according to this argument, it is necessary that whatever is placed at the centre remain there, so that, if fire were placed at the centre, it would be stationary there. So this is the way in which he responds to the doctrine: agreeing, to start with, that uniformity is the explanation, he refutes the doctrine on the grounds that it ought to be the explanation of the same thing in the case of the other <elements>, and then he objects that it is not necessary that uniformity be the explanation of resting, but that rather the explanation is that which is also the explanation of motion towards the centre; for <earth> is observed not only to remain at the centre, but also to move to it. And next, since the whole earth is not perceived moving to the centre, he argues for this with the words, 'for it is necessary that the whole <earth> move wherever any part of it moves' – since it is homoiomerous. However, a thing remains naturally
533,1 where it moves naturally – for this was proved previously.[343] So, if being uniformly related to the extremities is also common to the other elements, and moving naturally to the centre is unique to earth, the explanation <of its resting at the centre> would be not uniformity but its kinship with the centre, because of which it moves to the centre and remains there.

(295b25) He says that those who inquire about the earth's being at rest also do this strange thing: they inquire concerning it 'why the earth remains at the centre, but not why fire remains at the extremity'. For the word 'remains' applies in common to both subjects.[344] For this would be a path of discovery for them in the case of the earth. For if they found that this remaining at rest at the extremity belongs to fire by nature, it would be clear that it is necessary that there also be by nature some place for earth in which it remains at rest. But if a position at the extremity is not fire's by nature, the centre is not earth's by nature, and, if this is so, it would be necessary to make the necessity of uniformity responsible.

However, this account is fictional and similar to the one about the hair. The sophists say that if a hair which is homoiomerous were stretched strongly and the stretching were uniform through the whole of it, it would not break; for why should it break in this part rather than that, since the hair is similar in all its parts and the stretching of it is uniform? Similarly in the case of a person who is extremely hungry and thirsty, each equally, and equally in need of food and drink (and therefore yearning <for each> equally), he also necessarily remains stationary and does not move in either direction; for why should he first move in this direction but not in that one, since the need and desire are equal? So, if the places are not by nature, but the earth remains at rest because of the necessity of uniformity (which is similar to the fictions we have just mentioned), they should inquire about the resting above of fire. For if it does not remain at rest by nature, the theory causes difficulty and causes even more difficulty than in the case of the earth, since it is not possible to make uniformity responsible in the case of <fire>. But if <fire rests> by nature, the same explanation will apply to earth.

His furnishing of the examples of uniformity with mocking elegance is wonderful. But it is clear that the hair will break, since to hypothesise that it is homoiomerous in this way is a fiction. But also the uniform stretching of the extremities and the centre is impossible. In the case of the other example, even if the person is equally distant <from food and drink> the thirst will be more urgent. But even if one <desire> is no more urgent than the other, he will choose whichever he happens to choose, since when two equally pleasant dishes are presented to us we choose[345] first whichever we happen to. For uniformity does not completely prevent choice, although uniformity does make one's approach slower because it is drawn to the alternative.

(296a1) He says that it is also surprising that they inquire about the earth's being at rest but not about its motion. <They do not ask> for what reason one thing moves upward, another towards the centre, if nothing prevents them. For the discovery of the explanation for remaining at rest follows from discovery of it in the case of motion, and conversely (since they coincide).

(296a3) He has previously proved that in the case of the earth's being at rest the theory of uniformity is not necessary because another explanation is easier, that in terms of nature, since it is clear that, if <the earth> moves towards the centre naturally, it also remains there naturally. And now he proves that what they say is not even true at all *per se*, although it is true in the indirect sense that 'it is necessary that anything for which it is no more suitable to move here than there remain at the centre'. For if something remains at the centre naturally, it is necessary for it to remain at the centre, and if it is necessary for something to remain at the centre, then it has the feature of not moving more in one direction than another. So, even if it is true to say that it is necessary for that 'for which it is no more suitable to move here than there' to remain at the centre, this is not because that is the explanation of its remaining, but because that is attached to what by necessity remains.

(296a6) Since this is not the explanation of <the earth's> remaining at rest, he shows the extent to which on this theory there is nothing to prevent motion of the following kind: not indeed together as a whole, but being torn apart.[346] For also, if fire were placed in the centre, it would necessarily remain at the centre as far as the theory of uniformity is concerned, since it would have the same relation to all the extremity, which is homoiomerous. But we see that fire does not remain, but it moves in every direction from the centre to the extremity, if nothing prevents it, 'except that it <does not move> as a whole towards one point' but a proportional part of the fire goes to a proportional part of the extremity; for example, a fourth part of the fire goes to a fourth part of the surrounding <extremity> (it is possible for it to be divided proportionally since no body is without parts, since <only> the point shows <partlessness>). He says that what follows necessarily from uniformity is only that the body will not move to one point, but not that it will not move.

(296a17) He has said that as far as uniformity is concerned, just as if one hypothesised that fire is lying at the centre, nothing would prevent its moving in its parts because of uniformity, so too in the case of the earth. He now brings in another kind of motion. He says 535,1 that it would be possible, since uniformity does not prevent this kind of motion at all, for the earth to be rarefied and become greater from being smaller and to move from the centre if this place did not belong to it by nature so that it does not grow and become very large.

This, then, is what Aristotle says, and one might suppose that in fact Aristotle is not referring to the Platonic demonstration at all. For Plato makes even balance and uniformity responsible, but Aristotle obviously does not mention even balance anywhere (even if Alexander always understands even balance as equivalent to uniformity). How-

ever, even if Aristotle is directing his discussion at Plato, it should be said that Socrates carries out his discussion on the basis of the hypothesis that if the earth is in the centre of the heaven and is round, it will need nothing to keep it from falling, not air and not a vortex.[347] For natural even balance is not a matter of chance; rather it is what draws the earth, which is homoiomerous, uniformly to the central point, and the homoiomerousness and even balance of the heaven are sufficient for keeping the earth at rest. For it will not ever depart voluntarily either as a whole or in its parts; for why should it go here rather than there, since it is uniformly related to the whole heaven? But it won't go in its parts since this place belongs to it naturally. Nor will the heaven ever force it, either in its parts (since the whole heaven is homoiomerous with itself) or as a whole (since it is always the same and always has the same relation to the earth). But fire, even if it is, like the earth, homoiomerous and evenly balanced, is not evenly balanced relative to the central point, but evenly balanced relative to the perimeter. Therefore, when fire is placed at the central point, a position which is foreign to it, it strives to ascend to its proper place[348] by the briefest path, and in fact if someone placed all of fire at the central point, it would be necessary that it be torn apart; and earth would undergo the same thing if it were placed at the perimeter. But, unless some of it changed, fire wouldn't move as a whole or in its parts from the perimeter either, since it is natural for it and heaven is related uniformly to it; rather it is also true to say of it that 'for a thing which is evenly balanced and placed near something uniform will not be able to incline more or less in any direction'.[349] But because fire is near heaven and more unified with it, after maintaining its own position as heaven does, it moves along with heaven, whereas earth, being akin to the central point, remains around that. Notice that Socrates hypothesises what Aristotle says is the explanation of the earth's being at rest, <namely> the fact that this position is natural for it, and he adds the physical cause because of which, when things are in their proper places they do not need some other 536,1
superfluity such as air or a vortex in order not to move from there, since even balance and the uniformity of what contains them is sufficient for them. But Aristotle is dealing with those who think that uniformity and even balance is sufficient for there being no motion even in the case of things lying in a place which is foreign to them, so that he tries to refute the argument by setting out the example of fire which is placed here, and he everywhere brings the cause around to being in a place naturally, considering this principle to be sufficient. But Socrates has also provided the explanation of this. And it seems to me that something of the same kind results for Aristotle in connection with the principles of the elements, since he is satisfied with the qualities hot and dry and their opposites, but Timaeus <Locrus> also

searches for the principles of these things and ascends to the figures.[350]

Having finished up his historical account of the shape, position, and rest of the earth, he next turns to his own opinions about these things.

[Chapter 14]

296a24-b26 Let us first say whether it moves or remains at rest, [since as we have said,[351] some people make it one of the stars, and others, who place it at the centre, say it winds and moves around the central pole. That this is impossible is clear if one takes as a starting point that, whether it is away from the centre or at it, if it moves it is necessary that it be moved by force. For the motion doesn't belong to the earth itself, since, if it did, each of its parts would have this motion; but, in fact, they all move in a straight line towards the centre. Therefore the <alleged motion of the earth> could not be eternal, since it is forced and unnatural, but the order of the cosmos is eternal.

296a34 Furthermore everything that moves in a circle, except for the first <sphere>, is observed to be left behind and to have more than one motion, so that it would be necessary that the earth have two motions, whether it moved about the centre or while lying at the centre; and if this happened, it would be necessary that there be passage and turnings of the fastened stars. But this is not observed to happen; rather they always rise and set at the same places on the <earth>.

296b6 Furthermore, the natural motion of parts of the earth and of the whole is to the centre of the universe – this is why it now lies at the central point.

296b9 Since the centre of both is the same, one might raise the following difficulty: towards which do things having weight and the parts of the earth naturally move? Is it because it is the centre of the universe or of the earth? In fact it is necessary that it be towards the centre of the universe, since light things and fire, which move in the opposite direction as things with weight, move towards the extremity of the place which surrounds the centre.

296b15 But it happens to be the case that the centre of the earth and of the universe are the same. For they also move to the centre of the earth, but in an indirect sense insofar as it has its centre at the centre of the universe. A sign that they also move towards the centre of the earth is that weights, which move towards it, do not move in parallel but at similar angles so that they move towards one centre, which is also the centre of the earth.

296b21 It is evident then that it is necessary that earth be at the centre and motionless both for the reasons given and because weights which are projected upward by force in a straight line return to the same <place> even if the power propels them to infinity.

296b25 So that it neither moves nor lies away from the centre] is evident from these things.

He first proposes to prove that the earth does not move in a circle either around the centre as one of the stars, as the Pythagoreans said, or in the centre winding around the axis of the universe. And he proves that it is impossible that it move in this way using the second hypothetical mode[352] as follows:

(i) If the earth moves in a circle, it is necessary that it be moved by force;
(ii) but it is impossible for it to be moved by force and unnaturally;
(iii) therefore it is impossible for the earth to move in a circle.

And he proves the conditional as follows:

(i) If the earth is not moved in a circle by force, it moves naturally, since what moves is either moved by force or naturally;
(ii) but if it moved in a circle naturally, each of its parts would have this motion;
(iii) however, its parts do not move in a circle – rather they move in a straight line towards the centre;
(iv) therefore, the earth does not move in a circle naturally;
(v) therefore, it is moved by force.

And he tacitly proves the opposite, which says that it is not moved by force, categorically in the second figure as follows:

(i) The earth is eternal;
(ii) what is moved by force and unnaturally is not eternal;
(iii) therefore the earth is not moved by force.

He proves that the earth is eternal on the basis of the fact that the order of the cosmos is eternal, so that the earth is also eternal.

(296a34) Next he adds this second argument. If the earth moves in 537,1
a circle, whether about the centre or at the centre, it is necessary that it have two motions, just as all the other <stars> which come after the fixed sphere have the motion of the fixed sphere and their own motion around the poles of the zodiac, which is in reverse to the fixed sphere,

as a result of which they are left behind by the fixed sphere. For if any one of the planets rises today with the star on the heart of Leo,[353] say, when the planet moves, say, two degrees towards the following signs so that it comes to be at the fifth degree of Leo, the planet is observed to be left behind, since the star on the heart subsequently rises and sets before it. So if the earth moved in a circle, it is clear that it too would have its own motion opposite to that of the fixed sphere, so that it would be left behind by the fixed sphere; and furthermore it would, in fact, move around the poles of the zodiac just like the planets, which are more honourable than it. But if this were so, the fixed stars would no longer be observed by us to rise and set at the same points on the horizon as they do now, but instead the planets would, since they would be moving around the same poles as the earth; but the complete opposite is observed. And next, if the earth moved at the same speed as some one of the planets, the planet would be seen at the same point, neither rising or setting, which is clearly unreasonable. (One should notice that it has been indicated <by Aristotle> that, if the alleged circular motion of the earth were to be like the other circular motions around the fixed sphere, the earth would also move around the poles of the zodiac; and <this is clear> from the anomaly he adduces, since the differences of the risings and settings do not follow simply from the motion of the earth or from being left behind, unless the earth also were to move around the poles of the zodiac or some poles other than those of the equator.)

(296b6) He adds a third argument based on the following kind of axiom which he has assumed previously:[354] something remains at rest naturally in a place to which it moves naturally. So if the natural motion of parts of earth and of the whole of it is towards the centre of the universe, and something remains at rest naturally in a place to which it moves naturally, then it remains at rest at the centre of the universe. And also fire which moves towards the perimeter remains at rest there, insofar as it is able to. Perhaps, too, he is indicating that the motion of the earth is towards the centre, not around it, since the motion of its parts and of the whole of it is to the centre of the universe.

(296b9) Having said that the motion of the earth is to the centre of the universe, he adds that, even if the earth does not move, as some
538,1 say, but lies at the central point, one might raise this difficulty. Since the centre of both the universe and the earth is the same thing in subtratum,[355] towards which centre 'do things having weight and the parts of the earth naturally move? Is it because it is the centre of the universe or of the earth?'. Having raised this difficulty he says that it is necessary for them to move towards the centre of the universe, and he proves this by reference to fire and light things in general. For these things – for example, fire – which move in the opposite direction to things with weight, being opposite to them, do not move to what is

up in themselves, but to what is up in the universe. Consequently, too, earth, moving in the opposite direction to fire and towards the opposite, would move to what is down in the universe and the centre of it.

Alexander says that the words 'move towards the extremity of the place which surrounds the centre' mean the same as '<move> towards the extremity of the universe and what is up, the heaven, by which what is down and the centre, the sublunary realm, are surrounded'. But perhaps <Aristotle> means by 'extremity of the place which surrounds the centre' the highest part of the air, to which fire moves, so that the place surrounding the centre would be the limit of the air on the side of the earth and its extremity would be its upper part into which both the purest air and fire move.

(296b15) So, having proved that heavy things move towards the central point of the universe, he adds that they also move towards the central point of the earth, and he proves this on the basis of the fact that weights do not move in parallel. For if they fell in parallel, they would not both converge to the same central point. He proves that they do not move in parallel but converge towards the central point on the basis of the fact that they move at similar angles. (Those who put angles under <the category of> quality called equal angles similar.)[356] Obviously angles are equal or similar when each is right, since:

> When a straight line set up on a straight line makes the adjacent angles equal to each other, each of the equal angles is right,

as we learned in the *Elements*.[357] And that weights naturally fall to earth at right angles is made clear by the fact that a column, wall, or other weight does not stand if it doesn't make a right angle with the plane.[358] And we have learned that <straight lines> which fall at right angles converge towards the central point from the 19th theorem of the third book of the *Elements* of which the enunciation is:

> If a straight line is tangent to a circle and a straight line is drawn from the point of contact at right angles[359] to the tangent, the central point of the circle will be on the line which has been drawn.

And the theorem before this proves the same thing; its enunciation is:

> If a straight line is tangent to a circle and a straight line is joined from the centre to the point of contact, the straight line which has been joined will be perpendicular to the tangent.

But since it has been proved in the *Elements* that the central point is

539,1 on a straight line which is perpendicular to a tangent and a straight line through the central point is perpendicular to the tangent, but it is now proposed to prove that a straight line which is perpendicular to the plane of the earth converges to the central point, it should be shown that a straight line which is perpendicular to the tangent is also perpendicular to the plane of the earth:

Let ABC be a great circle in the surface of the earth, and let the straight line DE be drawn tangent to the circle <at A>, and let there be some weight F moving to the tangent on the straight line FA in such a way that the angles FAD, FAE are equal to one another. I say that the whole angle FAB is equal to the whole angle FAC. But these angles are produced by the falling weight relative to the plane of the earth.

Let FA be extended to the central point G of the circle (for it has been proved that the central point is on the straight line perpendicular to the tangent). Since then the angles DAG, EAG are equal (since each is right) and the angle GAC is equal to the angle GAB (since they are angles of semicircles of the circle ABC),[360] therefore the remainders, the horn-shaped[361] angles at A <EAC, DAB> are equal to one another. But the angles FAD, FAE are also equal. Therefore, the whole angle FAC is equal to FAB and they are angles relative to the surface of the earth.

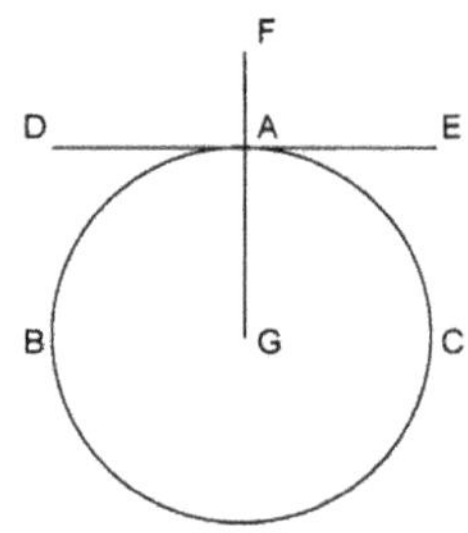

Figure 5 (539,5-540,4)[362]

At the start of the *Suntaxis*[363] Ptolemy hypothesises that weights fall to the earth at equal angles and therefore also remain uninclined in any direction and do not fall away, and he proves that they all press to the central point. For suppose FA stands at equal angles, like a kind of pillar. If we join a straight line from the central point <G> to A and we erect a straight line DE from A and at right angles to the diameter AG, FA and AG will be in a straight line; for since the whole angle FAB is assumed to be equal to FAC, and in them the one horn-shaped angle is equal to the other, the remainder FAD will be equal to FAE. Therefore each is a right angle. But DAG, EAG are also right angles. Therefore FAD, DAG are two right angles. Therefore,

FAG is one straight line by the 14th proposition of the first book of the *Elements*, of which this is the enunciation: 540,1

> If two straight lines not lying on the same side of a straight line make with the line at the same point the adjacent angles equal to two right angles, <the>[364] straight lines will be in a straight line with one another.

Therefore the weight F will move to the central point.

In general if all weights did not converge towards the central point, but moved in parallel, all of them would not move to the earth but some would move past the earth, since the earth, which is at the centre, does not extend over all the intermediate <region>.[365]

Therefore, if the earth moves to the centre naturally, it also remains at the centre naturally.

(296b21) He next adds another clearer demonstration that the earth doesn't move at all, either around the centre or away from the centre.[366] For if weights which are projected upward by force fall again to the same place in a straight line, even if the power projects them upward to infinity, so that there is a great amount of time between their projection upward and their fall, then the earth does not move. For, if the earth moved and the parts of it changed from place to place, what was projected upward would fall to a different place and not to that from which it was projected upward.

(289b25) Having resolved both of the issues and proved that the earth is at the centre and is motionless, neither moving in a circle around the centre or in a straight line, he adds the general conclusion, saying that it is evident from what has been said that 'it neither moves nor lies away from the centre', which is the same as 'it lies in the centre and is stationary'.

> **296b26-297a2** In addition, the cause of <the earth's> being at rest is clear from what has been said. [For if it is of such a nature as to move naturally to the centre from every place, as it is observed to do, and fire in turn moves from the centre to the extremity, it is impossible that any part of the earth move from the centre unless it is forced. For one thing has one motion and a simple thing has a simple motion, but not opposite motions. However, motion from the centre is opposite to motion to the centre. If, then, it is impossible for any part to move from the centre, it is evident that it is even more impossible for the whole to do so. For if the part is of such a nature as to move somewhere, the whole is also of such a nature as to move there. Consequently, since it is impossible for it to move except because of

> something with greater strength,] it is necessary that it remain[367] at the centre.

Having proved that the earth is stationary at the centre, he next explains the reason why it is stationary: it is not because of underlying air or the vortex, but because it is natural for it to remain at the centre. The argument is this. If its parts move naturally to the centre from every place, as they are observed to do, then, because fire also moves from the centre to the extremity, 'it is impossible that any part of the earth move from the centre unless it is forced'. For it is not possible that it move naturally both from the centre and to the centre because one body has one natural motion and a simple body has a
541,1 simple motion, 'but not opposite motions. However, motion from the centre is opposite to motion to the centre'. Consequently, if motion to the centre is natural, motion from the centre will not be natural, but, if there is such motion, it is by force. So if it is impossible for any part to move from the centre naturally, it is more impossible for the whole to do so. So if <earth> moves to the centre naturally, it is impossible that it move from the centre naturally. But if this is impossible, it is necessary that it remain at the centre. But the antecedent, the assertion that the parts of earth move towards the centre from every direction naturally, is true; therefore the consequent, the assertion that it remains at the centre naturally, is also true. This, then, is the reason it remains at the centre, that it is of a nature to do so. And one should not seek another reason.

> **297a2-8** What the mathematicians say in astronomy also bears witness to these things. [For the phenomena, the changing of the figures by which the order of the stars is determined, occur as if the earth lay at the centre.
>
> 297a6 About the manner of the earth's position and rest and motion] let this much be said.

He introduces as a witness to the fact that the earth lies at the centre and is stationary the fact that the astronomers also prove that the heavenly phenomena occur as if the earth lies at the centre and is stationary. What is said next is this. The phenomena, 'the changing of the figures by which the order of the stars is determined', occur as if the earth lay at the centre. For the order of the stars is determined with respect to the configurations which result from the motions. For if the earth were not in the centre of the heaven, the sizes of the stars, which appear in the east and then come to the meridian and finally reach the west, would not always appear to be equal. But if the earth had been displaced to the west the same stars would appear to be greater in setting than in rising, and if to the east, the opposite, and similarly for the intervals between them. And eclipses of the moon

would not have an order of their configurations such that they are always produced when the moon is in the position diametrically opposite the sun,[368] but they would also occur in intervals less than a semicircle. And this would happen if the earth changed place. But if it moved in a circle about the central point, as Heraclides of Pontus hypothesised,[369] and heavenly things were stationary, if it moved towards the west, the stars would be observed rising from there. And if it moved towards the east, then, if it rotated around the poles of the equator, the sun and the other planets would not rise at different places on the horizon, and, if it rotated around the poles of the zodiac, the fixed stars would not always rise from the same places as they now do; and whether it moved around the poles of the equator or around the poles of the zodiac, how would the change of place of the 542,1
planets into successive zodiacal signs be preserved, if heavenly things were motionless? Ptolemy also says this: <if the earth revolved from west to east> all things not standing <on the earth> would always be observed to have one motion opposite to that of the earth, and neither would a cloud ever be shown proceeding towards the east nor would any bird or projectile, since the earth would always exceed everything in speed, since it makes so great a revolution in such a brief time.[370]

(297a6) Having finished up what he has to say about the position of the earth (it is in the centre) and its rest and motion (it remains at rest and does not move), he turns next to the discussion of its shape.

297a8-b18 It is necessary that it have a spherical shape. [For each of its parts has weight as far as the centre, and when a smaller part is pushed by a greater it cannot swell, but rather one is compressed together by the other and yields to it until it comes to the centre.

297a12 One should understand what I say as if <the earth> comes to be in the way in which some of the physicists say it came to be, except that they make force responsible for the downward motion; but it is better to set out what is true and say that this happens because what has weight has the nature to move towards the centre. So, the mixture existing in potentiality, the things which were divided moved from every place towards the centre in a similar way. The result would be the same whether the parts which were drawn together at the centre from the extremities were similarly divided or arranged in some other way. So it is evident that, since they move from the extremities in a similar way from every direction towards a single centre, the bulk must become similar in every direction; for when an equal <amount> is added in every direction, it is necessary that the extremity be at an equal distance from the centre, but that is the shape of a sphere. And it will make no difference to the argument if the parts of the earth did not move together towards the centre in a similar way from every direction, since it is necessary that the greater part always push a smaller one in front of it forward since both have the same inclination up to the centre, and the heavier thing pushes the lesser weight forward up to the centre.

297a30 The difficulty which someone might raise has the same solution as these things: the earth being at the centre and spherical, if a weight many times as great were to be added to one hemisphere, the centre of the universe and of the earth would not be the same, so that the earth would either not remain at the centre or, if it did, it could also be stationary now, not occupying the centre, <in a position> in which it is its nature to move.[371]

297b1 So this is the difficulty, but it is not difficult to comprehend if we stretch a little and determine the way in which we think that any magnitude whatsoever moves towards the centre if it has weight. For it is clear that it does not do so until its extremity touches the central point; rather the greater magnitude must dominate until it takes the centre with its centre, since it retains its inclination until it reaches there.

297b7 Now there is no difference between saying this of a chance clod or piece of earth and saying it of the whole earth since what has been asserted does not follow[372] because of smallness or largeness, but with respect to everything with an inclination to the centre. Consequently if the earth moves from somewhere as a whole or in its parts, it is necessary that it move up to the point where[373] it assumes the centre equally from every direction and the smaller <parts> are equalised by the greater because of the forward thrust of its inclination.

297b14 So if <the earth> has come to be, it is necessary that it has come to be in this way, so that it is evident that its coming to be would be spherical, and, if it has not come to be but endures forever, remaining at rest, it is necessary that it be in the same way as it would have first come to be if it had come to be.] According to this argument then[374] it is necessary that its shape be spherical.

He proves that the earth is spherical through five arguments. He demonstrates the first, assuming at the start that the parts of earth are so constituted as to move as far as the centre – he indicates this with the words 'has weight as far as the centre', since what has weight and moves because of weight moves as far as the centre of the universe if it is not prevented. And all the parts incline towards the centre and the smaller ones are pushed and compressed by the greater because the greater always have a stronger inclination towards the centre; and when some are displaced by others from every direction the result is a spherical figure, just as, in the case of wax which is molded into a sphere, what extrudes is always pushed towards the centre.

Next he corrects our imagination, which places one part on top of another up to a great height until it loses its striving to the centre and goes somewhere else. For he says it does not swell, that is, it does not increase in bulk so that it spills over, since neither what is pushing nor what is pushed has an inclination towards anything else than the centre. Or perhaps he means by 'it cannot swell' that the smaller withdraws and changes places with the greater, since this is what happens to liquid things. But in the case of parts of the earth the smaller is instead compressed by the larger and the one is pressed together by the other or yields to it until the heavier comes to the centre, but, although they increase in bulk, they do not spill over as waves do.

(297a12) In order to clarify what he has said he hypothesises that the earth has come to be, as some of the physicists do (they indicate the ordering of primary and posterior things in the cosmos through their coming to be). However, these people make the vortex responsi- 543,1
ble, saying that the earth is moved towards the centre by force; but it is better and truer to say that this happens because what has weight has a nature to move towards the centre. On the hypothesis (which is thought to have been Anaxagoras') that the earth was mixed with other things previously and then divided out, coming to exist actually from existing potentially and moving to the centre, if heavy things were carried from the extremities equally from every direction to the centre (which is one), it would be necessary, since an equal <amount> is added from every direction, that 'the extremity be at an equal distance from the centre', but that is a spherical shape. And even if 'the parts of the earth did not move together towards the centre in a similar way from every direction', but more came from one direction, less from another, the same thing would result. And in this way it is necessary that the shape come to be spherical. For to the extent that the addition becomes greater, to that extent, with the weight increasing, what is underneath, which is smaller and lighter, would be pushed out by what is heavier until the lighter things pushed out by the heavier ones to the other side of the central point were collected together in an even balance with what was pushing them out and came to have a similar striving for the central point.

It is in this way that Alexander thinks that a weight which is pushed will not go beyond the central point. But what departs upward from the central point obviously leaves some room and what pushes it pushes it upward. But then how can what is heavier be pushed out upward naturally? For in general what has come to the central point will not still have weight so that it can push. And Aristotle clearly says that heavier things push as far as the centre, not that they push beyond the centre. So what he is saying is that the greater <weight> pushes the lesser forward and does this as far as the centre. The pushing forward then results when one thing pushes and compresses another of the things which yield (as he says) and increase in bulk because of the compression, since in this way it is put inside one surface. But also what continues to extrude and is further away from the central point strives to get nearer to it and moves down to a more hollow place and in this way a sphere is produced.[375]

(297a30) And so, having shown that the earth is made spherical using the hypothetical coming to be of the earth and that its centre is equally distant in every direction from the limits, he adds a difficulty which grows up out of what are called problems concerning centres of gravity by those who do mechanics. Archimedes and many others have written many very elegant treatises on centres of gravity. The treatises have as their purpose showing how one finds the central point of a given weight, that is, a certain point in the body such that if a cord is fastened to the point and the body is raised up, it will not incline in any direction.[376] And it is clear that the centre of a magnitude is not always the same as its centre of gravity.

The difficulty is this.[377] The earth being at the centre and being spherical, if some 'weight many times as great were to be added to one hemisphere', it is clear that, after the earth had been increased on one side by a magnitude many times as great as its original one, the centre of the earth would not be the same as it was before when the centre of the earth and that of the cosmos were the same. Consequently at that time the earth would not be stationary at the centre, since, having been increased by so much on one side, it would no longer be in the centre of the universe, or, if it were stationary when it was this way (this is what is meant by the phrase in commas 'or, if it did,') it could also be stationary now and lie outside the centre of the universe and not have the centre of the universe as its centre, even if no weight had been added to it on one side; for it is being hypothesised that it is stationary even when it has come to be outside the centre because of the addition. Having said 'it could also be stationary now, not occupying the centre', he adds, '<in a position> in which it is its nature to move'. He means that it could be stationary relative to that position outside the centre relative to which it is its nature not to be stationary but to move. But it is absurd that it be stationary in the place relative to which it is its nature to move. (He hypothesises that the added weight is many times as great <as the earth> so that the present centre of the earth's weight is very far away from the centre.)[378]

(297b1) He has raised the following difficulty based on the hypothesised addition taking place in the coming to be of the earth: if the earth had been increased on one side by a magnitude many times as

great as itself, it would be necessary, since its centre of gravity has been changed, that it incline to that side to which the addition of the weight was made and no longer remain at the centre, or if it were possible for it to remain at rest when it was outside the centre because of the addition, it could be stationary now apart from the addition, occupying something other than the centre, something relative to which it is its nature not to remain at rest but to move. But this is absurd, since nothing is stationary in a place in which it is its nature to move, but <only> in a place which it is not its nature to move from. And he now says, 'so this is the difficulty, but it is not difficult to comprehend' and recognise the difficulty and its solution 'if we stretch a little'. Obviously he means 'stretch our mind a little', or rather he means, 'stretch the previous argument' (the one hypothesising the coming to be of the earth) and specify the way in which we think that any magnitude whatsoever which has weight moves towards the centre. For before it was said in an indeterminate way that what has weight has a nature to move towards the centre. But now let it be specified that what has weight does not move towards the central point just as far as its own extremity touches the central point and, as it were, stands on it; but the greater weight always dominates over the lesser and either pushes it out or compresses it, and, uniting with it, it moves as far towards the centre as it can. Stretching and forming 545,1
a sphere with its own centre, that is with its concave surface, it encompasses the centre of the universe, since it retains its inclination until it reaches there. Consequently, even if a weight were added to the earth, as the argument raising the difficulty hypothesised, it would not remain with an extrusion, but it would also become spherical and encompass the surface of the earth, and again the original centre would also be the centre <of the earth> with the addition. For it is necessary that this earthy thing want to draw itself near to the centre, 'since it retains its inclination until it reaches there'. For if it continued to move and went beyond the centre of the universe with its centre, it would be moving upward and unnaturally.

(297b7) He says that there is no difference if one makes the argument for parts of the earth or the whole of it, since what has been asserted does not result because of smallness or largeness but because of a natural inclination to the centre. 'Consequently if the earth moves from somewhere as a whole' (as those who assert that the earth was at one time divided out from the mixture and carried to the centre) 'or in its parts, it is necessary that it move up to the point where', being spread around 'equally from every direction, it assumes the centre',[379] the smaller parts always taking on a surface which is equal to <that of> the greater ones. For since the greater and heavier move downward more, the smaller are pushed out upward by force by the forward thrust of this inclination which belongs to each of the greater, and they take on a surface equal to theirs. (Alexander says that some manuscripts say ‘it is necessary that it move with this’, that is, with an inclination to the centre, ‘until it assumes …’ in place of ‘it is necessary that it move up to the point where it assumes’. Alexander accepts the latter text.)[380]

(297b14) Finishing up the argument, Aristotle says that if the earth has come to be it is necessary that it has come to be in the way which has been described, in accordance with its inclination towards the centre, so that its shape is spherical, and if, being eternal, it endures, it is in the same way as it would have first come to be had it come to be. So in this way too it is necessary that the earth be spherical.

> **297b18-23** And <it is evident that it neither moves nor lies away from the centre> because all weights[381] move in similar angles [and not in parallel; but this belongs by nature <to what moves> towards what is by nature spherical. So either <the earth> is spherical, or, at least, it is by nature spherical. But one should speak about each thing in terms of the way it tends to be and exist by nature] and not[382] in terms of the way it tends to be and exist by force and unnaturally.

This is the second argument which proves that the shape of the earth is spherical on the basis of the fact that weights move in similar, that is, equal, angles. For when everything converges from every direction to the same place and one thing is always added to another with a similar inclination, it is necessary that the <straight lines drawn> to 546,1 the extremities from the centre to which everything converges be equal, and this <resulting> shape is spherical. Therefore, that weights fall in similar angles and not in parallel belongs by nature <to what moves> towards what is by nature spherical. For moving weights cannot move from every direction in similar angles unless that towards which they move has a spherical surface. But since the spherical shape of the earth was not made precise because of the extrusions of the mountains and its hollows, he says that either the earth is spherical (if one takes into account the fact that a difference of this sort is extremely slight, since, if someone put a millet seed on a huge wax sphere, one would not hesitate to call it a sphere because of the protrusion of the millet seed) or, if someone is speaking precisely, he will say that it is at least by nature spherical, since by nature it converges towards the centre from every direction, as is made clear by the fact that all weights fall at similar angles. So if one should speak about each thing in terms of how it tends to be and exist by nature and ‘not in terms of the way it tends to be and exist by force and unnaturally’, it is clear that it would be reasonable to say that the earth is spherical.

Alexander adds in a graceful way the reason why the shape of the earth is not precisely spherical, making anhomoiomerousness and lack of even balance responsible. He says:

> For it is not true that in the case of every heavy body the centre of inclination and gravity is precisely the centre of the magnitude as well; rather in some cases there is a difference. For all things with weight are not equally heavy, but things with weight do strive to occupy the centre with the centre of their own inclination, not the centre of their magnitude. Therefore, nothing prevents it being the case that, although the centre of the earth, that is, the centre of inclination, is in the centre of the universe, the distances of the magnitude of the earth from the centre are not equal in every direction.

> **297b23-30** <It is also possible to show that the earth is spherical> by means of perceptual phenomena, [since <if it were not> lunar eclipses would not have sections of the kind they do. For now in its monthly configurations it takes on all divisions (since it becomes straight and doubly convex and concave), but in the case of eclipses the delimiting line is always convex, so that, since it is eclipsed because of the interposition of the earth, the cause of the shape <of the delimiting line> will be the circumference of the earth], which is spherical.[383]

This third, clear argument is based on what appears to perception in a lunar eclipse. For, if the earth were not spherical, eclipses of the moon would not always be observed to have a convex line delimiting the illuminated <part> of the moon and the <part> covered in shadow. For if a lunar eclipse occurs when the moon falls into the shadow produced by the earth, it is necessary that the shadow be either conical or cylindrical or basket-shaped, since if it were of another shape, it would not always make the section of the lunar sphere convex. But all convex shadow shapes have been proved to be produced by a spherical shape. If the sun is greater than the earth, the shadow will be cone-shaped; if both are equal, it will be cylinder-shaped, and if the sun is smaller, it will be basket-shaped and like a truncated cone having its narrower <part> faced towards the earth.[384] So if the delimiting line is always convex and a line which is always convex is produced by some shadow[385] of this kind, and all shadows of this kind come from a spherical body, the earth will have a spherical shape. For if the earth were drum-shaped it would produce a shadow of this kind when the sun was in a certain position relative to the earth; but it would not always do so for every position unless it were spherical.

He reminds us that the fact that the delimiting line in eclipses is always convex does not result because the moon is not of such a nature as to be divided into sections by a straight line, but results from the shadow on the basis of the fact that in its monthly configurations (in which the shadow does nothing to it) the moon takes on all divisions, since the division becomes straight in its bisected phases, doubly convex in the doubly convex ones and concave in the crescent-shaped ones, but in eclipses it always has a convex dividing line.

> **297b30-298a15** Furthermore, it is evident from the appearance of the stars [not only that it <the earth> is round, but also that its size is not great. For when we change our position a little to the south or north the horizon obviously becomes different, the stars overhead change greatly and do not appear the same to those who move to the north or south; for some stars are seen in Egypt and in the vicinity of Cyprus but are not seen in the areas to the north, and the stars which are always seen in areas to the north are set in those regions. So, from these things it is clear that the shape of the earth is round and also is that of a sphere which is not large, since <if it were not, the difference> would not be so quickly obvious to those who change their position so slightly.
>
> 298a9 Therefore, those who assume that the region around the Herculean Pillars is connected to the region around India and in such a way that there is one sea do not seem to assume things which are overly unbelievable. They say, using elephants as evidence, that their species occurs in both extreme regions, thinking that because they are connected the extreme regions] have this feature with each other.

This is the fourth argument demonstrating that the earth is spherical; it also demonstrates that its size has the relation of a point to the heaven. For if when we move to the north the star called Canopus is not observed at all but the stars of Ursa Major always seems to be overhead, but Canopus is observed on high by people in Theban Diospolis[386] and the last star of Ursa Major is observed by them to set and rise, then it is clear that horizons become different for people who change position slightly. But this happens because the earth is spherical. For if it were drum-shaped and we moved from the limits to the limits, the horizon would remain the same because we were progressing in one and the same plane, and the same things would always be observed as far as the position of the stars is concerned. So, if the horizons are different, the earth is spherical; and it is also not large, since the horizons change obviously even if we do not proceed far; for if a sphere is large, the difference of the horizons would not be easily seen straightaway.

(298a9) But, he says, if the earth is not very large, one should not consider that those who assume that the westernmost and eastern-

most regions which are known to us (that is, the region around 548,1
Gadeira[387] and the Pillars of Hercules (which he calls Herculean)[388] and the region around India) are connected with one another and not from far away (so that the sea which is called the Red Sea and the sea near us are one) say things which are unbelievable. They use as evidence that these regions do not stand very far apart the fact that the species of elephant is in both regions, which are extremities with respect to our location, thinking that these extremities have this feature of bearing elephants (the same ones being distributed here and there, as it seems) because they are connected with one another. For I do not think that he wishes to indicate a similarity between these regions with these words but proximity, since similarity could also attach to things at a great distance; for the atmospheric conditions are practically the same for those who live in the same parallel <even if they live far apart>.

> **298a15-20** And the mathematicians who try to calculate the size [of the circumference say that it is about forty myriads.[389] On the basis of this evidence, it is necessary not only that the bulk of the earth be spherical, but also that its size not be great in relation to the other stars.]

He also introduces this last justification for <the belief that> the earth is spherical and that it is not great in comparison with the size of the stars; it is based on the witness of the mathematicians. For they taught its measurements and said that it is spherical and not very large, since they inferred its size and expressed it as so and so many stades.[390] But they also demonstrated that the sun itself is close to one hundred and seventy times the earth in size,[391] although it appears to be a foot <in breadth>[392] because of the distance; so, if someone observed the earth from the sun, he would see it as having a breadth which was one one hundred and seventieth of a foot. But if some of the fixed stars are greater than the sun although they appear to be so much smaller because of the distance, the earth would also exhibit an extremely small size in comparison with them. (He says 'the *other* stars' because the Pythagoreans, speaking symbolically, said that the earth was also one of the stars.)[393] Indeed, if one were to compare the earth with the whole fixed sphere, it would be seen to be really unextended and to have to the sphere the relation of a central point in the strict sense. This is also clear from the fact that the central points of the instruments used in astronomy, the meteoroscope and the armillary sphere,[394] are analogous to the central point of the universe when they are placed anywhere on the earth.

Since Aristotle mentions the measurement of the earth and says 549,1
that its circumference is forty myriads, it would be good also to write down briefly the method of measurement of the earlier people be-

cause of those who doubt their wisdom. Using a diopter,[395] they took two fixed stars distant from one another by a distance of one degree, that is one three hundred and sixtieth part of a great circle in the fixed sphere. With a diopter they found the places[396] at which the two stars are at the zenith, and they measured the distance between the places with an odometer[397] and found it to be five hundred stades. From this it follows that a great circle of the earth has a perimeter of eighteen myriads, as Ptolemy calculates it in the *Geography*.[398] And, since it is proved by Archimedes that the perimeter of a circle is greater than three times its diameter by a seventh of the diameter,[399] the diameter of the earth will be five myriads plus seven thousand two hundred and seventy-three.[400] And again it is proved that the <rectangle> contained by the diameter and one fourth of the perimeter is equal to the area of the circle,[401] so that the plane of the circle is in stades twenty-five double myriads of area and seven thousand seven hundred and twenty-eight simple myriads and a further five thousand.[402] And again it is proved that the whole surface of a sphere is four times the area of a great circle <of the sphere>,[403] so that it follows that the size of the surface is in stades one hundred and three double myriads and nine hundred and fourteen <simple> myriads.[404] To find the size in three dimensions of the earth,[405] one multiplies a great circle, which has twenty-five double myriads and seven thousand seven hundred and twenty-eight simple myriads and a further five thousand monads, by the diameter, and the result is a cylinder which has a base equal to a great circle and a height equal to the diameter, <containing> one hundred and forty-seven triple myriads,[406] six thousand and eighty-eight double myriads, four thousand three hundred and eighty simple myriads and a further five thousand monads.[407] And since a cylinder of this kind is one and a half times the sphere, I subtract a third of this <number> and I have as the remainder the number of the volume of the sphere of the earth: ninety-eight triple myriads, four thousand and sixty-three double myriads, six thousand four hundred and forty-six simple myriads and a further nine thousand five hundred and three monads.[408] Seeing that the earth has this great a size, the heights of mountains are sufficient neither to do away with its spherical shape nor to destroy
550,1 the measurements which are inferred on the basis of the assumption that it is spherical. For Eratosthenes proved, using diopters[409] which measure from distances, that a straight line falling perpendicularly from the higher[410] mountains to the lowest places is ten stades <long>.

If Aristotle says that the mathematicians infer that the size of the circumference of the earth is about forty myriads (and he is certainly calling its surface its 'circumference'),[411] since he has not added that this is the measurement in terms of stades, it is unclear whether he is in disagreement with the number of stades in the surface of the earth which was inferred later. But it would not be surprising if he were in disagreement, since <in his time> the theorems provided by Archimedes for the steadfast apprehension of the present subject had not yet been discovered. But perhaps Aristotle does not clearly accept this measurement as precise, but he only takes this much from it: however great the measures of the earth are, it is not very large.

Notes

1. The sentence, which has been the subject of considerable controversy, is Eudemus text 146 Wehrli (1955). (471,1-9 is DK12A19.) See Hall (1971), which includes references to some earlier discussions of the passage. I take the sense of what Eudemus said to be that Anaximander offered an incorrect account of the sizes and distances of some of the stars and the Pythagoreans gave a 'correct' account of all of them. On Eudemus see Goulet (1989-), vol. 3, pp. 285-9.

2. Reading *parabolês* with D, E, F, and Karsten. A has *meta parabolês*. Heiberg prints the otherwise unattested word *metaparabolês* without explanation.

3. On Aristarchus see Heath (1913), part 2, on Hipparchus, *DSB*, vol. 15, pp. 207-22, and on Ptolemy, *DSB*, vol. 11, pp. 186-206.

4. The introduction of sidereal periods here is a potential source of confusion, which sometimes obscures the discussion which follows: to say that Saturn is restored more slowly than the moon is to speak of angular speeds, to say the moon has a slower east-west motion than Saturn is to speak in terms of linear speeds. The difference between the two kinds of speed becomes reasonably clear starting at 474,30.

5. cf. *Cael.* 2.8, 289b15-16 and 289b34-296a5.

6. Here Simplicius uses *enantios* in a loose way, since he accepts Aristotle's claim (*Cael.* 1.4) that there is no opposite to a circular motion.

7. The east-west motion would seem to be forced because it is a matter of being carried around by the fixed sphere; the west-east motion would seem to be forced in the sense that it is resisted by the east-west motion of the fixed sphere. Alexander responds by saying that the planet 'chooses' to be moved by the fixed sphere. Simplicius accepts this for the east-west motion (473,2-6), but not for the west-east one.

8. On the Aristotelian commentator Alexander of Aphrodisias, to whom Simplicius constantly refers in this commentary, see Sharples (1987).

9. See the note on 471,31.

10. At 2.12, 292a20-1 below.

11. In what follows Simplicius gives a Neoplatonist account of the phenomena Aristotle is considering. He imagines that planetary spheres 'process' from the fixed sphere and come to be less and less like it and more 'planetary' as they get smaller.

12. Heiberg's suggestion that there is a lacuna here does not seem compelling, but, given that we are dealing with a 'sentence' of 216 words, it is difficult to be certain.

13. It seems unlikely that Alexander would have said that the sidereal periods of Mars and Mercury are the same when that of Mars is practically twice as great as that of Mercury. Perhaps one should emend *Areos* in line 9 to *Hêliou*. In any case Simplicius only discusses the relative position of Mercury and

Venus, which have the same period (and the same as the sun), although on the geocentric hypothesis one of them has to be further away from the earth than the other. (Simplicius returns to this difficulty at 476,28-477,2, where he leaves it unresolved.) On the question of the relative order of Mercury and Venus see Neugebauer (1975), pp. 691-2. Simplicius espouses the standard order, adopted notably by Ptolemy, with Mercury below Venus, but Alexander's order had more supporters than Simplicius suggests. It is found, for example, in the *De Mundo* ascribed to Aristotle (392a16-31); for other references see also Bousset (1915), pp. 31-6. Alexander's claim is presumably that Mercury has a west-east motion which is faster than Venus, but that both are restored in an equal time, so that Venus is less hindered by the motion of the fixed sphere.

14. *Rep.* 10, 617A. Since Plato does not name the planets in this passage, it is an inference (albeit a common one) that the sixth planet is Venus rather than Mercury. For a defence of the view that the sixth planet is Venus see Bidez (1935).

15. This sentence is badly formulated, but its point is clear enough. In the *Hypotyposis* (Manitius (1909), 224,4-6) Proclus mentions the same observation in connection with unnamed individuals who carried out a calculation like that described by Simplicius in the next sentence: 'Venus is observed to move under Mars, just as Mercury is observed to move under Venus.' Theon of Smyrna (Hiller (1878), 193,13-20), using a theory in which Mercury and Venus revolve around the sun, makes clear the unreliability of any such observation: '<Mercury and Venus> sometimes seem to obstruct (*epiprosthein*) one another, coming to be above and beneath each other because of the sizes, obliquities, and positions of their circles. However, precision is not easy (*adêlos*) in their case because they revolve around the sun, and because Mercury, being a small point close to the sun and brightly illuminated, is especially difficult to observe (*aphanê*).'

16. Perhaps *apostaseôs* should be supplied here. What Simplicius is describing is similar to what Ptolemy does in the second part of book 1 of his *Planetary Hypotheses* (Goldstein (1967)), sections 3 and 4, transforming his models for planetary motion into a cosmological picture; for a brief description see Pedersen (1974), pp. 391-7. See also Proclus, *Hypotyposis* (Manitius (1909)), 221,16-224,26 and *in Tim.* (Diehl (1903-6)), vol. 3, 62,16-63,20, and Hartner (1964).

17. Now restoration is discussed in terms of linear speed.

18. i.e. west-east along the ecliptic.

19. i.e. east-west along the equator.

20. Translating the *iousan te kai kratoumenên* of Karsten. Heiberg prints *iousês* (all the MSS he reports have *ousês*) *te kai kratoumenês*, which is also the text of Rivaud (*Oeuv. Comp.*, vol. 10); on the difficulty of this reading see Taylor (1928), ad 39A1.

21. Heiberg prints *periietai* with A; Karsten and Rivaud (*Oeuv. Comp.*, vol. 10) print *periêiein*; F has *periêiei*.

22. *Tim.* 38E6-39A3.

23. i.e. outermost.

24. Simplicius omits the *ienai* of our Plato MSS.

25. Heiberg prints *deuteron*, Chambry (*Oeuv. Comp.*, vol. 7.2) and Karsten *deuterous*.

26. Simplicius omits the words *hôs sphisi phainesthai apanakukloumenon*.

27. *Rep.* 10, 617A8-B3.

28. That is, Plato is considering the angular, not the linear west-east speed of the planets.

29. That is, their west-east motion.
30. So that planetary spheres closer to the fixed stars are more hindered.
31. Simplicius refers to the cosmos which is finite and therefore has only finite power, but is able to continue moving forever because of the prime mover.
32. I do not know to whom Simplicius is referring.
33. That is, every heavenly sphere will move only in an east-west direction, and those closer to the fixed sphere will move faster than those further away.
34. That is, the planets would not appear to have a daily motion parallel to the equator.
35. The isodromic stars are Mercury, Venus, and the sun which have equal sidereal periods. Simplicius correctly points out that there is no way to account for this behaviour if all planets are assumed to revolve around the earth and speeds are supposed to be proportional to distance from the earth, whether one assumes that planets closer to the fixed sphere have a slower east-west motion or a faster west-east one than planets further away from it. Cf. 474,7-13 with the note.
36. Simplicius cites *Cael.* 2.8, 290a7-9. For his text see the first note on the lemma at 452,7 in the commentary on ch. 8.
37. See the next lemma.
38. Cf. 290a31 in ch. 8, where Aristotle says that nature does nothing by chance.
39. *Cael.* 2.8, 290a7-29.
40. 'They' seems clearly to be Alexander.
41. Simplicius is perfectly right to reject the solution he ascribes to Alexander. But his own position (apparently that if p and q are equivalent in some strong sense, each can be used to prove the other) is weak.
42. *Cael.* 2.8, 290a35-b7; for a textual issue see the first note on the lemma at 458,8 in the commentary on ch. 8.
43. Simplicius leaves out an Aristotelian *men gar*.
44. Translating the *de* (or *d'*) of F, D, E, and Karsten, which is the reading of all manuscripts of Aristotle. Heiberg prints *de, ei,* a correction of *de eis* in A.
45. *dikhomênon,* a word used by Aratus in the *Phaenomena* (Kidd (1997)) at lines 78, 471, and 737 to characterise the full moon, the mid-point of the lunar month; cf. Geminus (Aujac (1975)) 8.11. Simplicius is worried because in the course of a lunar month there are two half-moons and Aristotle speaks of the moon being only once bisected in the course of its waxing and waning. But he decides – correctly – to take 'bisected' as a characterisation of the half moon.
46. *Cael.* 2.12, 292a3-6.
47. I take this to be Alexander; cf. 477,27 with the note.
48. That is, faced away from us.
49. That is, when the straight lines from the sun and moon to the earth form a right angle.
50. There is a fuller description of the sun's illumination of the moon in Cleomedes (Todd (1990)), 2.5.41-80. I owe this reference and many significant corrections to Alan Bowen.
51. The drum-shaped case is equivalent to the case of a flat disc always faced towards the earth, the lentil-shaped case to a convex disc with a face always towards the earth. Simplicius is noticeably vague on the lentil shape. In any case the arguments about the apparent shapes of the moon and eclipses of the sun would not rule out that the moon is only approximately a sphere.
52. Translating Heiberg's suggestion *toiautas tas* instead of the *tas* which he prints.

53. Heiberg prints *aporiôn ousôn*, Moraux *aporiain ousain*, which is also found in D, E, F, and Karsten.

54. According to Schoch (1927), col. xx, the only possible date for this occultation is 4th May, 356.

55. In a citation of these first words at 481,21-2 Heiberg prints a *men* here which is not printed by Moraux or Karsten.

56. Heiberg prints *ouden an alogon einai doxeie*, and Moraux *ouden alogon an doxeien einai*, which is also printed by Karsten. It appears from Heiberg's apparatus that D has *ouden alogon an doxeien*, E *ouden alogon an doxeien* followed by one erased letter, and F *ouden an alogon doxeien einai*.

57. A question mark here is probably preferable to Heiberg's period.

58. *Tim.* 38C7-D2; see also *Rep.* 10, 616E8-617A1. Simplicius contrasts this order with the, for him, standard one of Ptolemy, in which the sun lies above Mercury and Venus.

59. Simplicius closely paraphrases Aristotle here, but he has *asterôn* where Aristotle has *astrôn*.

60. Here again Heiberg prints a period.

61. Simplicius pretty much quotes 292a14-15 but adds the verb 'attain' changing the *kai* of line 15 from an 'even' to an 'and'. Alexander proposed adding the verb 'demand' to the text. It seems clear that Simplicius is right to suggest that no addition is needed, but his own paraphrase beginning with 'Rather what he says' is not easy to reconcile with the text he quotes.

62. Simplicius refers to *Cael.* 2.3, 286a3-6, which I have translated, 'Since motion in a circle is not opposite to a motion in a circle, we should inquire why there is a plurality of motions, although we have to try to make the inquiry from far away, not far away spatially but much more because we perceive very few of their features.' I have translated the word 'features' (*sumbebêkotôn*) of this passage as 'what occurs' in the present text because it picks up on the phrase 'what occurs' (*sumbainontôn*) in the passage on which Simplicius is commenting.

63. Heiberg prints *monon autôn*, Moraux and Karsten *autôn monon*, which is also the reading of E; F has *autôn monôn*. Simplicius cites these words at 378,15 in the commentary on ch. 1 and 388,23 in the commentary on ch. 2, on which see the notes.

64. *koniseôs*. Simplicius offers his explanation of this problematic word at 483,1-2.

65. The meaning of this last phrase is explained in the next lemma at 292b17.

66. On the text here and its meaning see below 483,13-15 with the note.

67. On Simplicius' text here see 484,5 with the note.

68. Translating the *gar* of Simplicius' citation at 484,10. Aristotle and Karsten have *de*.

69. Moraux prints *aei estin*. Both words are missing from Simplicius' citation at 484,10 as printed by Heiberg (they are present in Karsten). I have translated Heiberg's text here.

70. At 484,16 Simplicius cites these words as *kai dê kai tôn allôn*, where Moraux prints *tôn d' allôn*.

71. Simplicius discusses these difficult words at 484,21-31.

72. Simplicius takes it that in 292a18-28 Aristotle treats the seven planets as things which achieve the good through several motions, and that he only addresses the difficulty starting at 292a28. See 483,7ff.

73. Reading the *ê* of A, C, D, E, F, and Karsten rather than the *hêi* which Heiberg prints.

74. i.e., the prime mover.

75. For the complicated textual situation in Aristotle see Moraux (1954), pp. 158-9; he believes that the alternative text proposed by Simplicius is not 'Coans' but 'Chians or Coans'. The die (*astragalos*) referred to here is a very rough six-sided object. For a photograph, see Jenkins (1986), ill. 36; and for drawings, Pottier and Reinach (1887), vol. 1, p. 217. The following description is based on Daremberg and Saglio (1875), vol. 5, p. 29: 'The only difference between the *astragalos* and the die (*kubos*) is that the former was elongated and could not stand on its two extremities, which are too thin and rounded. Of the other four faces a broad and slightly convex one was called the back (*pranês*); opposite it was the broad and slightly concave "belly" (*huptia*); the remaining two faces were long and narrow; one slightly hollowed out was called the Coan and was the least stable of all; opposite it was the Chian.' For the scholarly disagreement on which side was the Coan, which the Chian, see PW, vol. 13, col. 1934. I do not know the point of saying there are large dice on the two islands.

76. The standard Neoplatonic human goal, taken from Plato, *Theaetetus* 176B.

77. So Simplicius does not see Aristotle as providing a specific solution to the difficulty, but only pointing out that there are two types of hierarchies relative to action, one expressed in terms of the number of actions in a chain needed to achieve a good, the other in terms of the number of different types of actions one can perform. This double hierarchy makes it possible to explain the difference between the sun and moon and the other planets. The sun and moon may have fewer motions because they can achieve the good with fewer; the planets may have more because they can do more things.

78. Heiberg here prints *tôi de isôs arista ekhonti*, Karsten *tôi d' hôs arista ekhonti*, which is also printed by Moraux at 292b4-5. With the *hôs* it seems clear that Aristotle is talking about what is in the best condition, but it becomes difficult to see why Aristotle should say that what is in the best condition is that for the sake of which. Without the *hôs, to arista ekhon* can be translated either 'what is in the best condition' or 'being in the best condition'. It seems that in this paragraph Simplicius relies on this ambiguity. I have consistently translated *to arista ekhon* as 'what is in the best condition' although this translation undermines the plausibility of what Simplicius says.

79. cf. *GA*. 2.23, 731a24-6: 'The substance of plants has no other function or action than generating seed.'

80. Simplicius first considers severing 292b8-11 from the preceding reference to plants to avoid having Aristotle say that they might have a plurality of activities, but then suggests a way in which this might be true.

81. I have inserted these words in the light of Simplicius' paraphrase at 486,5 (which inserts an *autôi*).

82. Moraux prints this sentence as *alla mekhri hotou dunatai tukhein tês theiotatês arkhês*. Simplicius' 'citation' at 487,2-3 is *mekhri gar hou dunatai, phêsi, metekhei tês theiotatês arkhês*.

83. Heiberg prints *tôn kinêseôn* with A: Moraux prints *kinêseôn*, as does Karsten with D, E, and F.

84. See the discussion of the preceding lemma with the note on 484,2. It is difficult to see how Alexander's suggestion could solve the difficulty raised by Aristotle solely in terms of the number of steps needed to achieve the good.

85. This sentence is Aristotle fr. 49 (Rose (1886)).

86. Simplicius switches to the singular *oligê kinêsis* ('slight motion') from the plural *oligai kinêseis* ('few motions', sometimes 'slight motions'). I take him to be distinguishing between things which achieve an ultimate good with a single simple motion and things which can only perform a few of the motions which they would require to attain such a good and so only achieve a partial good.

87. I take Simplicius to be saying that what attains an ultimate end with many motions is a mean between what attains it with slight motion and what can only attain a partial end. This enables one to say that all the planets are better than sublunary things and not as good as the fixed sphere. This resolves the difficulty raised by Aristotle only if it is assumed that sun and moon don't attain the good attained by the higher planets, an option Simplicius adopts in the following paragraphs.

88. Reading the *auto* of Karsten rather than the *ep' auto* printed by Heiberg.

89. It seems reasonably clear that Simplicius has not found a fully satisfactory solution to the difficulty raised by Aristotle. He is unhappy with saying that the sun and moon do not attain the highest good at all (as sublunary things do not), and so he considers with apparent reluctance saying that they do not attain it in a fully perfect way, perhaps explicable as becoming identical with the good.

90. In this sentence Simplicius gives a formulation of the account of the creation of time in Plato's *Timaeus* (37C-38E). The 'perfection' is the world of Forms, the first heaven is the sphere of the fixed stars and endures forever, and time breaks up the unity of the world of Forms.

91. The myth is the story of the overthrow of the chief god *Ouranos* (heaven) by his son, the Titan *Kronos* (Saturn); see Hesiod, *Theogony* (West (1966), 154-82. In the interpretation followed by Simplicius, *Kronos* is identified with *khronos* (time). On this identification, which is very common, see PW, vol. 11, cols 1986-7, 2008-9. For *Kronos* as the source of division and particularisation see Proclus, *in Tim.* (Diehl (1903-6), vol. 3, 188,2-20). A reader gave me much help with this paragraph.

92. In this paragraph and elsewhere Simplicius uses astronomical terminology without explaining it. A *phasis* here is presumably a heliacal rising or first visibility, a notion normally only applied to the fixed stars, but applicable by extension to the five planets (although it is not clear why the moon could not be said to have *phaseis* in this sense as well). The word recurs with the same sense at 488,6 and 496,2; at 547,13 it is used for phases of the moon. In the course of their generally west-east motion through the zodiac, the five planets (but not the sun and moon) are observed to stand still (*stêrizein*) or to have stations (*stêrigmoi*) and to move in reverse. In astronomical terminology one zodiacal sign, say Taurus, which is to the east of another, say Aries, is said to follow (*hepesthai*) the other, because the sun is in Taurus after it is in Aries (see 421,20ff. in the commentary on ch. 5 and 496,22, 537,7, and 542,2); and Aries is said to precede (*proêgeisthai*) Taurus because the sun is in Aries before it is in Taurus (see 421,12.15.19 in the commentary on ch. 5). The west-east motion of a planet is said to be towards (*epi*) or into *(eis)* the following or successive signs (*ta hepomena* (*zôidia*)) (496,22; 537,7; 542,2); Simplicius describes the east-west motion of the lunar nodes as *epi ta proêgoumena* ('towards preceding signs') at 497,13. On the other hand, an east-west motion is called a *proêgêsis* (literally, 'motion forward'), which I have translated 'retrogression' because the *proêgesis* of a planet is a motion contrary to its west-east motion through the zodiac (here and at 488,6 and 504,28). Simplicius calls a west-east motion of a planet an *akolouthêsis*, which I have translated 'progression' (here and at 488,7). In the text here and at 488,6.10 Simplicius uses the noun *hupopodismos*, which I have translated more literally as 'backward motion' (cf. the use of the verb *hupopodizein* at 491,24), since it refers to the same thing as *proêgêsis*, but is understood in relation to a planet's west-east motion through the zodiac rather than in relation to the east-west motion of the fixed stars. Simplicius also uses the noun *propodismos* (488,5) and the verb *propodizein* (491,24), which I have translated 'forward motion' and 'move forward'.

93. i.e. that they change their speed.

94. On Sosigenes, the teacher of Alexander of Aphrodisias, see Moraux (1984), pp. 335-60. For a brief discussion of this complicated sentence, which is Eudemus text 148 (Wehrli (1955)) and Eudoxus fr. 121 (Lasserre (1966)), see Görgemanns (1970), p. 93. Simplicius discusses the astronomical theories of Eudoxus starting at 492,25.

95. Reading *planômenôn* or *planêtôn* for the *aplanôn* in Heiberg's text, of which I can make no sense. I note that a series of words printed by Heiberg do not occur in D or E, giving the sense 'as if there really were several motions'.

96. At 822A4-8 of the *Laws* the Athenian criticises the belief that the sun, moon, and other stars 'wander'. He continues, 'Each of them has the same path; they do not traverse many paths, but always just one, which is circular, although they appear to move through many.' But in the *Timaeus* (35E-40B), Timaeus makes the apparent motion of the planets a composite of a west-east motion and the east-west motion of the fixed stars. Some modern scholars (and apparently some ancients) have seen an inconsistency here, but others have not (see, for example, Vlastos (1975), pp. 101-2, who espouses a reconciliation quite like Simplicius' here).

97. I had first supposed that Simplicius was here referring to all animals, who are composed of relatively small portions of the elements, but I have followed a reader's suggestion that Simplicius is only thinking of very small creatures such as insects.

98. i.e. the world soul of Plato's *Timaeus*.

99. Simplicius is presumably thinking of *Metaph.* 12.7, 1072b26-30, where Aristotle is describing the prime mover; since this prime mover is intellect, it is also a part or kind of soul.

100. On the earth as hearth (*hestia*) of the universe see PW, vol. 8, cols 1293-8.

101. *arkhê*, which Simplicius paraphrases as sovereign worth (*arkhikon axiôma*) at 490,7 and then, it seems, as power (*dunamis*) at 490,9 and 14.

102. Simplicius picks up on the first formulation of the difficulty at 292a10-4.

103. For the issue, which relates to the breaking off of this lemma, see the discussion of the next lemma starting at 490,17.

104. Trying to translate the *eilêptai* of D, F, and Bessarion's correction of the *eileiptai* of E. Heiberg prints *eirêken* with A. Karsten prints *eilêphen*.

105. The prime mover.

106. 'This' is feminine in the Greek. In what follows Simplicius indicates that Alexander supplied a noun for 'this' in order to make the present lemma contain an independent response to the second difficulty. Simplicius seems uncertain what to do, but when he considers taking this to be the superiority, a more appropriate translation of these first words would be 'and this superiority will be proportional, ...'.

107. Again, at least Alexander; see 490,29-491,3.

108. i.e. in the previous lemma.

109. Heiberg prints *kai eti dia touto hen ekhousi sôma hai allai sphairai* with A and F. Moraux and Karsten print *kai eti dia tode* (*tod'* Karsten) *hen ekhousi sôma hai allai phorai*, which is the reading of D and perhaps E. (Heiberg's apparatus indicates that E has *tode*, but he says nothing about *sphairai*.) [Alexander] (*in Metaph.* (*CAG*, vol. 1), 703,2-4) cites the words as *kai eti dia touto hen ekhousi sôma kai hai allai phorai*. I have translated *sphairai* because there is no trace of *phorai* in Simplicius' discussion (although there is nothing which could be called a paraphrase of this first part of the lemma).

110. cf. *Cael.* 1.7, 274b33-275b4. The following final four lines of ch. 12 (293a11-15) are not included in any lemma and are not discussed by Simplicius: 'We have spoken about the stars which have a circular motion and said what they are like both in terms of their substance and in terms of their shape, and we have spoken about their motion and their order.'

111. See the discussion of the preceding lemma. 491,15-510,35 are translated into French in Aujac (1979), pp. 157-99, with a facing Greek text, a slightly modified version of Heiberg, from which I have taken several changes as indicated in subsequent notes.

112. cf. [Alexander], *in Metaph.* (*CAG*, vol. 1), 703,22-3. 491,17-28 are text 165B in Fortenbaugh et al. (1992). On Theophrastus see PW, suppl. vol. 7, cols 1354-1562.

113. The *idios* of D, printed in Aujac (1979), is preferable to the *idion* of A printed by Heiberg.

114. i.e. move faster and slower.

115. In *Cael.* 1.3, 269b18-270a12.

116. See 488,3-29.

117. At this point Simplicius begins a long excursus (which ends at 510,35) on the history of planetary theory, probably derived mainly, if not entirely, from Sosigenes, on whom see the note at 488,20; Simplicius does not mention Sosigenes again until 498,2. Moraux (1984), pp. 347-51, gives a useful philological analysis of the passage. Lasserre (1966) prints 492,31-497,8 as fr. 124 of Eudoxus.

118. See 488,18-24.

119. This clause has been variously interpreted and translated. I more or less follow the suggestion first made by Böckh (1863), p. 155, according to whom it says that Callippus, who was a pupil of Polemarchus, who in turn was a pupil of Eudoxus, came to Athens after Eudoxus, etc. Others (e.g. Dicks (1970), p. 190) suppose that Callippus came to Athens after Polemarchus. On Polemarchus, about whom our only other information comes from 505,18-23, see PW, vol. 21, cols 1256-8; on Callippus see PW, suppl. vol. 4, cols 1431-8.

120. At 488,18-24 Simplicius says that Eudoxus was the first to carry out this kind of investigation. If he has anyone specific in mind here, perhaps it is Pythagoras; see below 507,12-14. On the description of the theory of Eudoxus which follows see, e.g. Heath (1913), pp. 193-211, who summarises the fuller discussion of Schiaparelli (1874). Schiaparelli's influential account has been challenged frequently, most recently by Mendell (1998 and 2000) and Yavetz (1998 and 2001).

121. That Eudoxus believed in this latitudinal variation in the sun's west-east motion is attested by Hipparchus (Manitius (1894), 1.9.2, partly identical with Eudoxus fr. 63b Lasserre (1966)). Lasserre rejects the idea. Because the idea is problematic (see Neugebauer (1975), pp. 629-31), I postpone discussion of the role and speed of the second and third spheres until Simplicius takes up the case of the moon, starting at 494,23.

122. This is a careless formulation, since only the outer and middle spheres for the sun are starless; cf. 491,18-20. 493,17-20 are text 165C in Fortenbaugh et al. (1992).

123. This work has not survived, and is known only from the descriptions of Eudoxus' planetary system which we have (Lasserre (1966), fr. 121-6).

124. The *mesôn* of D, E, F, and Karsten, printed by Aujac (1979) is preferable to the *mesou* of A, printed by Heiberg.

125. That is, the angular distance between the planes of the second and third circles corresponds to the distance the moon departs from the ecliptic.

126. That is, the moon's greatest deviations in latitude (and therefore its intersections with the ecliptic (nodes)) occur further and further to the west in the zodiac; see the note on 487,25.

127. The *dusmas* of D, E, F, and Karsten, printed by Aujac (1979), is preferable to the *dusmôn* of A, printed by Heiberg.

128. Most commentators since Ideler (1830), p. 77, have assumed that Simplicius has interchanged the second and third sphere in the case of the moon and, presumably, of the sun, even though Aristotle (*Metaph.* 12.8, 1073b17-22) assigns the same position to the second and third spheres for sun and moon as Simplicius does (although Aristotle does not mention their speeds). I quote Heath (1913), p. 197, on the moon: 'The object of the third sphere was ... to account for the retrograde motion of the nodes in about $18^1/_2$ years. But it is clear (as Ideler saw) that Simplicius' statement about the speeds of the third and second sphere is incorrect. If it had been the third sphere which moves slowly, as he says, the moon would only have passed through each node once [in the course of the $18^1/_2$ years] and would have been found for nine years north, and then for nine years south, of the ecliptic. In order that the moon may pass through the nodes as often as it is observed to do, it is necessary to interchange the speeds of the second and third spheres as given by Simplicius; that is, we must assume that the third sphere produces the monthly revolution of the moon from west to east round a circle inclined to the ecliptic at an angle equal to the greatest latitude of the moon, and then that this oblique circle is carried round by the second sphere in a retrograde sense [in approximately $18^1/_2$ years].' The underlying assumption of this criticism is that Simplicius (or his source) is correct to say that Eudoxus was interested in explaining the retrogression of the lunar nodes. The assumption may be anachronistic; cf. Dicks (1970), pp. 180-1.

129. At *Metaph.* 12.8, 1073b22-32. But the description which follows is more detailed than Aristotle's. (*Metaph.* 12.8, 1073a17-1074a15 and the commentary on it traditionally ascribed to Alexander are translated in the Appendix.)

130. Venus.

131. On the association of Saturn with the sun (*Hêlios*) see Boll (1917-19). The values given here are standard, reasonable approximations; see e.g. Heath (1913), p. 208.

132. It is actually the equator of the second sphere.

133. Simplicius' description does not enable one to determine which of two possible rotations Eudoxus assigned to the third sphere, but this turns out to make no substantive difference.

134. See the note on 487,24.

135. *diexodou khronon*, i.e. the synodic period. I have found no occurrence of the term in this sense, although LSJ give three occurrences of *diexodos* meaning something like 'orbit' in Herodotus, Euripides' *Andromache*, and the *De Mundo* traditionally ascribed to Aristotle.

136. Or three and two-thirds months if we follow F and Karsten. As this variation suggests, months are taken to be thirty days long; cf. Heath (1913), pp. 285-6.

137. Except for the case of Mars, where the number given is about one third of the true value, these figures are approximately correct; cf. Heath (1913), p. 208, and, for further discussion, Mendell (1998), pp. 213-18, where possible emendations of the number for Mars are considered. The words 'according to what Eudoxus thought' (*kathaper Eudoxos ôieto*) suggest that Simplicius is aware of the discrepancy.

138. We have no quantitative information about the inclination of the poles of the fourth sphere to those of the third. For his reconstruction of Eudoxus' theory, Schiaparelli (1874) calculated inclinations using the now accepted lengths of the arcs of retrogradation of each of the planets. The notion of opposite rotation is, of course, problematic in the case of spheres with axes inclined to one another. Schiaparelli made sense of it by assuming that the inclination had to be less than 90 degrees, an assumption which led to quite unsatisfactory descriptions of the motions of some planets, most clearly for Mars. For an alternative understanding of opposite rotation see Yavetz (1998), pp. 231-2.

139. i.e. the equator of the third sphere.

140. *hippopedên*, normally just transliterated in this context. For a fuller description of this curve, roughly a figure eight lying on its side, see e.g. Heath (1913), pp. 202-7.

141. i.e. the ecliptic.

142. The words 'i.e. the order of their distances' are bracketed by Ross (1953) and Jaeger (1957).

143. Heiberg prints *melloi* where Karsten prints *mellei* in agreement with Ross (1953) and Jaeger (1957) and E.

144. Heiberg prints an *ana* here, which is not in our texts of Aristotle and not printed by Karsten.

145. *Metaph.* 12.8, 1073b32-8.

146. The next four sentences constitute Eudemus text 149 Wehrli (1955).

147. On Euctemon and Meton see *DSB*, vol. 4, pp. 459-60 and 9, 337-40. In the Eudoxan system the west-east motion of the sun is uniform, implying equality of the seasons. Callippus actually made a more accurate determination of the length of the seasons than Euctemon and Meton; see Heath (1913), pp. 215-16.

148. Presumably the additional spheres for the moon were also intended to account for its non-uniform west-east motion. Unfortunately, Simplicius does not tell us Callippus' reason, and Schiaparelli (1874), p. 101 (p. 187 in the German translation) supposed a lacuna here. For Schiaparelli's conjectures about the additional spheres for the five planets see Heath (1913), pp. 212-16, and see now Mendell (1998), pp. 229-60.

149. Heiberg prints *eis tauton apokathistôsas*, Ross (1953) and Jaeger (1957) *eis to auto apokathistasas*. F and Karsten have *eis tauto apokathistasas*.

150. *Metaph.* 12.8, 1073b38-1074a5.

151. Adding a *kai* to Heiberg's text with D, E, F, and Karsten; Heiberg follows A.

152. Simplicius' last mention of Sosigenes is at 510,24. Between here and there Simplicius quotes and/or paraphrases him extensively. Heiberg does not indicate that what follows is a quotation. Moraux (1984), p. 347, says that here 'scheint <Simplikios> wörtlich zu zitieren'.

153. Starting at 499,17.

154. Moraux (1984), p. 349, n. 68, takes the words which follow down to 499,17 to be Simplicius rather than Sosigenes.

155. *Metaph.* 12.8, 1074a2-4. The Heiberg text here and a few lines below is the same as that reported in the note on 498,1. Heiberg's apparatus indicates that only Karsten has *apokathistasas* in both places.

156. Aujac (1979) brackets this sentence as a badly inserted gloss.

157. Moraux (1984), p. 347, calls what follows up to 501,21 a resumé; he does not mention the *phêsin* at 501,1.

158. This figure, which is based on Schiaparelli (1874), fig. 20, and those in the following pages are not in our texts of Simplicius. For understanding the next two arguments it helps to visualise the 'initial' situation as one in which the points D and F are under A.

159. Translating the *kinoumenês eite menousês tês periekhousês ekeinas* of F (Karsten does not have the *ekeinas*). Heiberg prints *menousês eite perieilousês ekeinas*, noting that A has *periekhousês*, D and E *periagomenês*. Aujac (1979) prints *periagomenês eite menousês tês periekhousês*.

160. Translating the *menonta* of F and Karsten rather than the *menousa* of D, E, printed by Heiberg (A has *malista*).

161. Heiberg calls the text corrupt here. I have attempted to render the text he prints. In any case the sense is not in doubt.

162. The argument is that the effect of the motion of AB is to keep D and F under A, the effect of the motion of DE is to make D and F cease to be under A, and the counter effect of the motion of FG is to bring F back under A.

163. Sosigenes indicates that the motion of a point A along the diagonal AC of a parallelogram ABCD can be described as a motion of A along AB while AB moves down to coincide with CD when A reaches B.

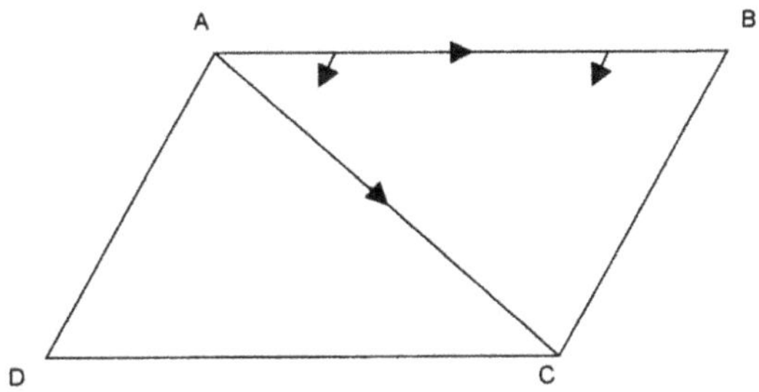

164. cf. Theodosius (Heiberg (1927)) 1,15.

165. Conjecturing *ton Dia [tettarôn]* for the *tên dia tettarôn* printed by Heiberg (D does have *ton* for *tên* here).

166. Schiaparelli (1874), p. 106 (German translation, p. 192), takes this very vague phrase (*kai meiôsei kata to phainomenon to takhos*) to be an inexact expression of the point that the number of spheres which would affect the motion of Jupiter is reduced by one. Aujac (1979) renders it 'elle en modifiera donc la vitesse pour qu' il y ait accord avec les faits observés'.

167. As it stands, this sentence is unsatisfactory. I have emended 'second' to 'seventh' and changed an imperfect 'added' to a present 'adds'. The second sphere by itself does not add to the east-west speed of the spheres under it, although one might say it does so in combination with the first sphere. The important point, which is brought out in my version of this sentence and certainly in the remainder of the paragraph (provided we assume some reference to the seventh sphere), is that in itself the seventh sphere moves from east to west along the ecliptic, but since with it the effect of the second, third, and fourth spheres are counteracted, it has as a composite motion the motion of the sphere of the fixed stars. This fact means that one really shouldn't need a new sphere representing the sphere of the fixed stars in the system for Jupiter. Simplicius goes on to discuss this point in the next paragraph; see also 503,36-504,3. At 506,23-507,8 he points out that one could get along with just the sphere of the fixed stars and no spheres representing it for the planets and hence no spheres counteracting such representing spheres.

168. Simplicius cites *Metaph.* 12.8, 1074a7-8. From here to 504,3 he is concerned with *Metaph.* 12.8, 1074a1-14.

169. The interpretation envisaged starts from five spheres for the sun, three of Eudoxus, two of Callippus; for these Aristotle would add four counteractive spheres. The interpretation proposes that he considers the result of subtracting these four spheres plus the two added by Callippus, plus the two added by Callippus for the moon. But, Simplicius objects, this leaves two Eudoxan solar spheres, the second and the third, uncounteracted.

170. That is, spheres counteracting the effect of the spheres of the sun on the spheres of the moon.

171. One would prefer the plural here, which Schiaparelli (1874), p. 108 (German translation, p. 194), and Aujac (1979) supply without textual warrant.

172. cf. [Alexander], *in Metaph.* (*CAG*, vol. 1), 705,39-706,15, translated in the Appendix.

173. Heiberg here prints *Metaphusikês*, an otherwise unattested word, with A rather than the standard *Meta ta phusika* of F and Karsten.

174. Lines 22-34 are Porphyry fr. 163 Smith (1993); on Porphyry see PW, vol. 22, cols 275-313.

175. i.e. the number should be 49 rather than 47.

176. That is, to adopt the proposal considered at 502,19-27, according to which the last counteractive sphere for Saturn (the seventh sphere) is the first sphere for Jupiter (the eighth sphere). If similar identifications are made for the planets below Jupiter, the number of spheres in the Aristotelian system would turn out to be 49, not 55.

177. This renders the *eu legôn* which Simplicius puts at the end of the following 'quotation', which expands on the criticism of the idea of making the seventh and eighth spheres identical. 504,4-15 are text 165D of Fortenbaugh et al. (1992).

178. There is, of course, something unsatisfactory about calling both the spheres of Eudoxus and Callippus and those added by Aristotle 'counteractive'.

179. Translating the *phoras* of Karsten rather than the *diaphoras* of A and F printed by Heiberg.

180. Most of the text from here to 506,3 is translated by Heath (1913), pp. 221-3. 504,16-22 is printed by Lasserre (1966) as fr. 126 of Eudoxus.

181. On Autolycus see *DSB*, vol. 1, pp. 338-9.

182. An otherwise unknown person, once thought to be a teacher of Aratus; see PW, vol. 2, cols 1055-6, with suppl. vol. 1, col. 136. Perhaps one should revive Schiaparelli's suggestion ((1874), p. 109, n. 1 (German translation, p. 109, n. *)) to read 'Aristotle'.

183. cf. Pliny, *HN* (Beaujeu (1950)) 2.6.37, Martianus Capella (Willis (1983)), 8.883, Isidore of Seville, *Liber de Natura Rerum* (Fontaine (1960)), 23.1.

184. *tôn autôn peri to di' hou theôreitai kathestôtôn.* Something like this seems to be the meaning; Heath (1913), p. 222, has 'under the same conditions as to medium', whereas Schiaparelli (1874), p. 109 (German translation, p. 195), offers a quite different rendering.

185. Literally, eleven-fingered and twelve-fingered. On the use of the 'finger' as a unit of measurement see Neugebauer (1975), pp. 658-9.

186. On the treatment of annular eclipses in antiquity see Görgemanns (1970), pp. 136-9. In the *Hypotyposis* (Manitius (1909), 130,16-23) Proclus apparently refers to this passage from Sosigenes.

187. i.e. change of distance from the earth.

188. This sort of expression is not used in extant standard astronomical texts (as opposed to mathematical commentaries and philosophical discussions) but Ptolemy uses it in his *Harmonics* (Düring (1930)), e.g. at 103,24.

189. Set by Plato (488,18-24).

190. cf. the note on 493,7.

191. This sentence is Aristotle, fr. 211 Rose (1886). Schiaparelli (1874), pp. 110-11 (German translation, p. 196), gives no credence to the suggestion that Aristotle had misgivings about homocentric spheres; such suggestions, he says, are due to the desire to excuse the later Peripatetics' abandonment of homocentric spheres in favour of eccentric circles and epicycles.

192. I have tried to translate the *autôi* of A rather than the *auton* of F (and Karsten) printed by Heiberg, which a reader suggests could be translated 'he was won over by <the thesis> that the spheres are homocentric etc.'.

193. Heiberg ends Simplicius' quotation of Sosigenes at this point.

194. Heiberg prints an *oun* here which is omitted in our texts of Aristotle and by Karsten.

195. Heiberg prints *hupolambanein*, Karsten and Aristotle have *hupolabein*.

196. Heiberg omits a *para* printed by Karsten and found in Aristotle.

197. *Metaph.* 12.8, 1073b11-17.

198. Heiberg omits an *oun* which is printed by Karsten and in Aristotle.

199. Heiberg prints *phorôn*, which is the reading of Themistius in his paraphrase of *Metaph.* 12 (*CAG*, vol. 5.5, 28,14), whereas the MSS of Aristotle and apparently [Alexander] (*CAG*, vol. 1, 706,18) have *sphairôn*, which is printed by Karsten. See Ross (1953) ad loc. for this and for the words 'and perceptible' which follow.

200. The words 'and perceptible' (*kai tas aisthêtas*) are found in all MSS of Aristotle, but bracketed by Ross (1953) and Jaeger (1957).

201. *Metaph.* 12.8, 1074a14-17.

202. On Callisthenes, the nephew of Aristotle, who accompanied Alexander the Great to the East, see Prandi (1985).

203. The Latin translation b has a more plausible 'one thousand nine hundred and three'.

204. This sentence is Porphyry fr. 164 (Smith (1993)), and part of it is *FGrH* 124.3. On this report and other parallel pieces of 'obvious nonsense' see Neugebauer (1975), pp. 608-10. See also Toomer (1988).

205. Simplicius cites book 2 of Ptolemy's *Planetary Hypotheses*, which does not survive in Greek, but is preserved in an Arabic translation, rendered into German by Ludwig Nix (Heiberg (1898-), vol. 2, 111-45). The Arabic text is printed by Goldstein (1967). In the next notes I cite parallel passages from Nix's translation.

206. 'Erstens weil die Dinge am Himmel nicht viele Bewegungen haben wegen des Verhaltens der Sphären, welche sich einander drehen, da es wohl möglich ist sich vorzustellen, dies geschehe in wenigen Bewegungen' (Heiberg (1898-), vol. 2, 117,21-4). I have not found anything in the text which refers specifically to simultaneous restoration.

207. 'Das wunderbarste hierbei ist aber dass sie die letzten Sphären die ersten bewegen lassen und die umschlossenen die sie umschliessenden, die mehrfach anomalistischen die einfachen, ganz im Gegensatz zur natürlichen Lehre' (Heiberg (1898-), vol. 2, 118,27-31).

208. I quote a segment of an extremely long sentence: '... der Anfang <der> Bewegung in der in <einem Tier> Lebenskraft ist, dann ein Impuls von dieser Lebenskraft eintritt, der sich dann in den Muskeln zieht, dann von den Muskeln in die Füsse beispielhalber ... und hier zu Ende kommt ...' (Heiberg (1898-), vol. 2, 119,26-31).

209. On Iamblichus see Dillon (1987). He associates Pythagoras with both eccentrics and epicycles in his life of Pythagoras (Deubner (1975), 6.31). For other such ascriptions to Pythagoras see Burkert (1972), p. 325, n. 10. I have found nothing in the extant work of Nicomachus, on whom see Dillon (1996), pp. 352-61.

210. Simplicius would have done better to express this sentence as follows: let the circle ABCD through the middle of the signs of the zodiac not be homocentric with the universe in the sense that it is no longer the central point of the zodiac E on which we say the eye (and the earth) is, but F, and ABCD is no longer homocentric with the universe but eccentric relative to it.

211. Reading *kentrou* for the *ekkentrou* printed by Heiberg and Karsten.

212. See *El.* 1, 16.

213. I have bracketed the *perigeiou* printed by Heiberg. Karsten has *prosgeiou*, which at least makes sense.

214. Again see *El.* 1, 16.

215. Here, for the first time, Simplicius applies the word 'uniform' (*homalês*) to an angle rather than to a motion and contrasts uniform angles with apparent angles. The terminology is unfortunate, but it is clear from Simplicius' examples what he has in mind.

216. The statement of the differences relies on *El.* 1, 32.

217. The *tôi* (masc.) of F and Karsten and Aujac (1979) is preferable to the *pros to* of A printed by Heiberg.

218. For this aspect of Ptolemy's lunar theory see Neugebauer (1975), pp. 84-8, or (less complex) *DSB*, vol. 11, pp. 192-4. In the theory the centre of the eccentric circle is not stationary.

219. At *Cael.* 292a10-14.

220. It appears that Simplicius thinks that in the theory of epicycles only the epicycle moves the star, presumably because there is no reason to think that the motion of the epicycle FGHK around E should affect the motion of FGHK around A (see Figure 3). However, he suggests a possibility for accommodating epicycles and eccentrics to 'equalisation' below at 510,8-15.

221. At 504,16-505,27.

222. For the view that the stars rotate around their centres see 454,23-456,27 in the commentary on ch. 8.

223. On its own this assertion would not apply to eccentric spheres, but at 510,19-23 Simplicius suggests that eccentric spheres can be fitted inside <pairs of> homocentric spheres.

224. If 'he' is Aristotle, Simplicius is presumably referring to 293a3-4.

225. Simplicius here describes the planetary picture put forward by Ptolemy in book 2 of the *Planetary Hypotheses* (Heiberg (1898-), vol. 2, 123-45); for discussion see Neugebauer (1975), pp. 923-6.

226. Simplicius at 511,16 substitutes *hopou* for Aristotle's *hou*.

227. On this astronomical system, which is now usually ascribed to Philolaus, see Huffman (1993), pp. 231-61.

228. See 366,16-22 in the outline of book 2 at the beginning of the commentary on book 2.

229. In ch. 3 at 286a18-20.

230. In ch. 8 at 289b5-6.

231. I take Simplicius to be referring to ch. 4, 287a30-b4; see his commentary on that material, starting at 414,20.

232. cf. ch. 4, 287a30-b4.

233. Part of this sentence is included in DK59A88 (Anaxagoras).

234. 511,25-31 are part of Aristotle fr. 204 Rose (1886).

235. cf. Alexander, *in Metaph.* (*CAG*, vol. 1), 40,26-41,2.

236. *Zênos purgon*. Proclus, *in Tim.* (Diehl (1903-6), vol. 2, 106,22 and vol. 3, 141,11-2) and Simplicius, *in Phys.* (*CAG*, vol. 10, 1355,9) have *Zanos*.

237. *Dios phulakên.* Proclus (*in Euc.* (Friedlein (1873), 90,18) and *in Tim.* (Diehl (1903-6), vol. 2, 106,22)) has *Zanos.*

238. cf. the next lemma.

239. This sentence is a part of fr. 204 Rose (1886). The expression 'throne of Zeus' is not uncommon (it is used by Aeschylus, Sophocles, and Euripides), but I know no other identification of it with cosmic fire.

240. Reading the *ho* which is bracketed by Guthrie, Moraux, and Leggatt, following Allan (1936), but is retained by Longo (1961). Without the *ho* the words 'the centre <and>' would be dropped from my translation. Simplicius clearly read the *ho,* and at 513,12-32 he offers two interpretations of what Aristotle is saying. On the first, fire is the most authoritative part of the universe and is guarded at the centre; on the second, which Simplicius finally opts for at 514,21-3, the centre is the most authoritative part and is guarded by fire.

241. Simplicius replaces Aristotle's singular with a plural.

242. Heiberg prints *hôste kai topos hupo selênên ho mesos topos.* I have translated Karsten's *hôste kai tôn hupo selênên ho mesos topos timiôtatos.*

243. Of Tarsus, a Stoic of the late second century BC. This passage is *SVF* 16 for Archedemus (von Arnim (1903-24), vol. 3, p. 264). On the disputed reliability of this report, see PW, suppl. vol. 12, cols 1366-8.

244. As Heiberg indicates, the reference is to the present lemma.

245. Reading *prosêkei* with F and Karsten rather than the *prosêkein* printed by Heiberg following A.

246. Reading *tautên tên* with A, F, and Karsten rather than the *tên autên* substituted by Heiberg.

247. There is no grammatical break between the end of this lemma and the beginning of the next, and the next lemma raises an issue about the notion of centre.

248. I am inclined to read the *legôn* of F and Karsten rather than the *legei* printed by Heiberg on the basis of b.

249. Simplicius quotes from *Tim.* 40B8-C3, a sentence which he quotes in full at 517,7-9.

250. That is, because we are on the side of the earth away from the counterearths. On this paragraph and the next see Dicks (1970), pp. 66-70.

251. In the lemma Heiberg prints *apekhontôn hêmôn tês diametrou* with A. *hêmôn* is omitted in F, and Moraux does not print it (Karsten ends his lemma at 293b21).

252. In this passage Aristotle broaches and answers a question which is still open: did Plato in the *Timaeus* ascribe any kind of motion to the earth? (For some bibliography see Cherniss (1959-60), pp. 223-4.) The centre of the controversy is a word used by Plato at 40B8, which in the MSS is either *illomenên, heillomenên,* or *eillomenên.* The exact sense of these words is unclear. It does seem clear that Aristotle read *illomenên* and understood it to imply a rotation around the axis of the cosmos. I have translated *illomenên* as 'wound'. For Simplicius the word indicates that the earth is fixed, 'bound' as he puts it at 517,14.

253. Heiberg, following A, does not print a *men,* which is in F and Karsten and our MSS of Plato.

254. The *tên* of our MSS of Plato is missing in Simplicius. On this *tên* see Cornford (1937), p. 120, n. 1.

255. Karsten and our MSS of Plato have *polon tetamenon.* Heiberg prints *tetamenon polon.*

256. *Tim.* 40B8-C3.

257. In Simplicius' time, as in ours, a distinction was made between a pole (*polos*) and an axis (*axôn*), which, as Simplicius shows, was not always made earlier.

258. The word *polos* does not occur in our texts of the *Phaedo*. At *Axiochus* 371B2 it has the sense of heaven. In its other two occurrences in the Platonic corpus (*Cratylus* 405C9, *Epinomis* 986C4) it has the sense of heavenly or stellar revolution.

259. *Argonautica* (Fränkel (1961)), 1.129.

260. *illasin dêsantes*. Simplicius cites *Iliad* (West (2000)), 13.572.

261. See Snell et al. (1971-), vol. 3, p. 140 (Aeschylus, fr. 25).

262. *Phaedo* 109A4-5. This text is also quoted at 532,1-2 and 535,28-9, on which see the notes. Here Heiberg prints *klithênai* with the text of Plato. A has *kai theinai*, F and Karsten *kinêthênai*. At 532,2 Heiberg prints *klithênai* with F and Karsten, where A has *ekklithênai*, C *kinêthênai*.

263. The next two quotations are from section 31 of *On the Nature of the Kosmos and the Soul* of Timaeus Locrus (Marg (1972)). For Simplicius' use of Timaeus Locrus see Marg (1972), pp. 102-10.

264. Simplicius and the MSS of Timaeus Locrus have *horos*, which Marg (1972) prints as the Doric *ôros*.

265. Translating the *aous* of F and Karsten (Marg (1972) prints *aôs*). A has *agos*; Heiberg prints *augas*.

266. Marg prints *anatolas*, which is also the reading of F and Karsten. Heiberg prints *antolas*.

267. The words *hôs tâi opsei kai ta apotomai tas gâs perigraphometha* (which we describe by vision and the section of the earth), which Marg prints, are missing in Simplicius. For the plural 'horizons' see Baltes (1972), p. 108.

268. Reading *to strephomenês sêmainousês* with F and Karsten rather than the *sêmainousês* printed by Heiberg.

269. Reading *eirêmenon* with F and Karsten rather than the *eirêmenou* printed by Heiberg.

270. Reading *sustrephomenên* with F rather than the *strephomenên* printed by Heiberg; see 518,13, which makes clear that 'rolled up' was offered as a paraphrase of the reading *eillomenên* in the *Timaeus* passage.

271. Reading Karsten's *rhêseis* instead of the *khrêseis* printed by Heiberg.

272. i.e. as *eilomenên* (or perhaps *heilomenên*), which means the same thing as *eillomenên* (or *heillomenên*).

273. In the *Phaedo*, 109A.

274. Reading the *oun* of A, F, and Karsten rather than the *an* conjectured by Heiberg, and placing Karsten's period rather than Heiberg's comma after *ekdekhomenous*.

275. Heiberg prints *tautên* with A rather than the *tên gên* of F and Karsten. Either way the meaning is the same.

276. Simplicius refers ahead to the beginning of ch. 14, where Aristotle argues against the two hypotheses simultaneously.

277. As Moraux, p. clxv points out, this is one of the few cases in which Simplicius suggests an alteration of the text. According to Moraux only two manuscripts adopt Simplicius' suggestion.

278. Simplicius cites 293a15-16.

279. Simplicius refers ahead to the next lemma.

280. On Heraclides see the note on 541,28.

281. Heiberg's lemma here is unusual in only containing the initial words of the text and not its conclusion. Karsten's lemma has the standard form, and

Heiberg's apparatus ('Lemma ad p. 294a11 continuat F') indicates that F does as well.

282. For Simplicius' understanding of these words (which differs from that of, e.g., Leggatt) see 519,23-8.

283. Simplicius paraphrases proposition 22 of Euclid's *Optics* (Heiberg and Menge (1883-1916), vol. 7).

284. cf. 294b13-23 (below 524,1).

285. Simplicius glosses over the fact that at the end of the lemma Aristotle reverts to the topic of the earth's motion and rest.

286. Simplicius discusses this sentence and Alexander's handling of it at 521,14-27. For the sentence Moraux prints: *to de tas peri toutou luseis mê mallon atopous einai dokein tês aporias, thaumaseien an tis.* Simplicius' discussion is not clear and the MSS show considerable variation at various places involving *peri tas* and *tas peri* as well as *toutou* and *toutôn*. See my notes below and the very helpful discussion of Moraux (1954), pp. 159-61.

287. Simplicius describes material following the present lemma.

288. This sentence is part of DK59A88 (Anaxagoras).

289. Reading *peri tas toutôn* with Karsten rather than the *peri tas peri toutou* printed by Heiberg, following A, or the *tas peri toutou* of F and the Aldine. Moraux (1954), pp. 159-61, who takes the present sentence to represent Alexander's account of Aristotle's meaning, thinks that the correct text here would be *peri tas toutou* (the solutions of this).

290. Simplicius inserts the word *phulaxasthai*. Moraux ((1954), p. 160) assumes the insertion is Alexander's.

291. Reading the *toutôn* of F and Karsten rather than the *toutou* of A, printed by Heiberg.

292. Simplicius has not said this but he is apparently referring to the insertion of *phulaxasthai* at 521,15.

293. And we do not have to add *phulaxasthai*.

294. Reading *toutôn* with F and Karsten rather than the *toutou* of A, printed by Heiberg, and reading the *toutôn* in line with what Simplicius goes on to say.

295. Moraux prints a text which includes the words *ep' apeiron autên errizôsthai legontes* (saying that the earth is rooted *ad infinitum*). It seems reasonably clear from what Simplicius says at 522,5-11 that he did not have these words, since he expresses uncertainty about the referent of 'what is under the earth' and makes no mention of the unusual word *errhizôsthai*, which implies that the earth stretches down infinitely.

296. This paragraph and Simplicius' comment on it (522,7-12) are part of DK21A47 (Xenophanes); cf. DK21B28.

297. These lines constitute DK31B39; however, obeying the edict of Willamowitz-Moellendorff (1930), p. 249, DK substitute *glôssas elthonta* (coming through the tongues) for the *glôssês rhêthenta* printed by Moraux.

298. This paragraph and Simplicius' comment (522,14-18) constitute DK11A14.

299. *pheretai eis buthon*. At 522,29-31 Simplicius paraphrases this with *eis ton puthmena tou hudatos elthêi*, and suggests that he is offering a paraphrase of *buthizetai*.

300. This is a fairly common expression in commentaries on the logical works. Here it is presumably applied to *modus tollens*; [Ammonius], *in An. Pr.* (*CAG*, vol. 4.6), 68,25-9, gives as an example of conversion with antithesis 'if human, then animal; but not animal; therefore not human'.

301. *Top.* 1.18, 108a38-b6.

302. In *Top.* 1.18.

303. The text here causes difficulty. Moraux prints *athroôs* [*tôi*] *katôthen* (remains stationary as a mass under the earth). It is clear from 524,12-16 that Simplicius had the *athroôi tôi* of some MSS of Aristotle, but not clear exactly what he made of it. I have done the best I could with Simplicius' text.

304. The lemma up to this point is part of DK13A20 (Anaximenes); it is also mentioned in DK59A88 (Anaxagoras). On the clepsydra, which Simplicius describes at 524,17-525,4, see Last (1924).

305. This word appears to be a *hapax*.

306. Simplicius gives a second way of reading Aristotle's invocation of clepsydras in the next sentence.

307. Reading *eisienai aera* with F and Karsten rather than the *eisienai* printed by Heiberg.

308. I have translated *enseisantos auta* instead of the *enseisthentos autois* printed by Heiberg.

309. The reference is to *Cael.* 4.4, 311b6-13, where Aristotle apparently says that an inflated wineskin weighs more than an uninflated one.

310. The modern Khabur, which flows from southeast Turkey into Syria, where it joins the Euphrates. The Khabur is about 100 kilometers via an ancient trade route from Harran, the town where some think Simplicius wrote his commentaries; see Hadot (1987), pp. 9-21 (and in English in Hadot (1990)), for discussion and references, and see Lameer (1997) for sceptical remarks. Here Simplicius refers to the kelek, a raft which floats on inflated skins; see Tardieu (1990), pp. 71-95.

311. Moraux prints *to skhêma tês gês*, as does Karsten. Heiberg prints *to skhêma tautês* with A; F has *tês gês to skhêma*.

312. *tis*, printed by Moraux, but not by Heiberg in the paraphrase of this passage at 526,14, although it is included in F and by Karsten.

313. On this and the same expression just below see the note on 526,17.

314. *Cael.* 295a9-14 is included in DK59A88 (Anaxagoras), which also includes an edited version of 511,23-5 and 520,28-31.

315. For a textual issue here see the note on 527,18.

316. Translating the *aneirêke* of F, Karsten, and a correction in a second hand of A, which has *an eirêke*; Heiberg conjectures *anêirêke*.

317. *Phys.* 5.6 and *Cael.* 1.2.

318. A pun?

319. The *epizountes* in Heiberg's text is apparently a misprinting of *epizêtountes*. I have translated the *tên aitian zêtountes* of F and Karsten.

320. cf. the preceding lemma (294b13-14).

321. Translating *tou ouranou* with F and Karsten rather than the *toutou* printed by Heiberg.

322. *ekstantos*, Simplicius' paraphrase of the word I have translated 'yield' at 295a22. There, Moraux conjectures *hupeikontos* in place of the well-attested *hupelthontos*, for which Verdenius (1969), p. 280, provides a plausible defence.

323. Moraux prints *eipeien*. F and Karsten have *an eipoi*, A *eipoien*, C *eipoi*. Heiberg prints *eipoi an*.

324. For the evidence that Simplicius is correct to say that, according to Empedocles, the world is now under the domination of Strife see Wright (1981), pp. 45-8.

325. Alexander's second alternative involves interpreting *dieistêkei khôris hupo tou neikous* as something like *dieistêkei khôris tou neikous*.

326. Simplicius substitutes an aorist for an imperfect.

327. Simplicius now quotes the first 15 lines of DK31B35; he quotes all but the first two lines of this fragment at *in Phys.* (*CAG*, vol. 9), 32,12-33,2; for other partial citations see DK. For commentary see Wright (1981), pp. 206-8.

328. DK print the *logou* of Bergk (1842), col. 1002 (= Bergk (1886), p. 47), as does Karsten. Heiberg prints *logôi* with A and F.

329. Reading (with Wright (1981)) the unaccented *pô* of F and Karsten (which Heiberg accents); DK conjecture *tôn*.

330. DK31B86 (this passage the only source). The 'which' are presumably 'elements' which make up the eyes. In connection with this and the next two quotations it is useful to consult Theophrastus, *De Sensibus* 8-9 (Stratton (1917), pp. 72-4).

331. DK31B87 (this passage the only source). Again the 'them' are presumably the 'elements' which make up the eyes.

332. i.e., Aphrodite. This is DK31B95, for which this is the only source. Again the 'they' are presumably the 'elements' which make up the eyes.

333. Translating the *toss'* printed by DK and Karsten rather than the *toi'* of Stein (1852), p. 51, printed by Heiberg (*toia* in A and F).

334. DK31B71, for which this passage is our only source.

335. Translating the *eidea poipnuousa* of F and Karsten printed by DK rather than the *aither' epipneiousa* of Stein (1852), p. 58, printed by Heiberg. A has *ei de apopnoiousa*.

336. DK31B73, for which this is our only source.

337. DK31B75. Simplicius cites the last half again at *in Phys.* (*CAG*, vol. 9), 331,9 as an example of Empedocles' invocation of chance.

338. These lines are DK31B17, 7-8. Simplicius quotes all but one of the 35 lines of this fragment at *in Phys.* (*CAG*, vol. 9), 158,1-159,4, the missing line being supplied from *Phys.* 8.1, 250b30; for other citations see DK.

339. C, F, and Karsten add the words 'and Democritus' here, but A, which Heiberg follows, does not. Anaximander, of course, called the starting point of cosmogony *apeiron*, and the doxographers standardly applied the word to Anaximenes' starting point, air; see especially Simplicius, *in Phys.* (*CAG*, vol. 9), 22,9-13, where Simplicius says that the notion of infinity in magnitude applies to Anaximander and Anaximenes, who hypothesised that there was one element infinite in magnitude.

340. *Phaedo* 109A4-5, quoted earlier at 517,20-2 and below at 535,28-9, on which see the notes. In all three passages Heiberg prints *oudamose* with the text of Plato, F, and Karsten. A and C have *oudamôs* here, and A has it at 535,22.

341. See *Cael.* 293b30-2 with Simplicius' discussion at 517,1-519,8.

342. On this anomalous second explanation (perhaps explicable by the fact that Anaximander thought the earth was (more or less) flat) see Kahn (1960), p. 55, where the shape of Anaximander's earth is also discussed.

343. cf. *Cael.* 1.8, 276a22-4.

344. Simplicius indicates that one must supply the word 'remains' as indicated by the brackets in the translation of the lemma.

345. Reading *hairoumetha* with F and Karsten rather than the *anairoumetha* of A printed by Heiberg. b has *eligemus* ('we choose').

346. Heiberg prints *diaspômenon*, Karsten *diespasmenon*. For the corresponding Aristotle text Moraux prints *diespasmenon*; some Aristotle manuscripts have *diaspômenon*.

347. Heiberg prints *dinêseôs*, conjectured on the basis of the *dianoêsei* of A. F and Karsten have the perfectly intelligible *dinês*.

348. Reading the *ton* printed by Karsten rather than Heiberg's *to*.

349. *Phaedo* 109A4-5, quoted earlier at 517,20-2 and 532,1-2, on which see the notes. Here Heiberg (and Karsten) prints *plêsion* in place of the *en mesôi* of the text of Plato.

350. Simplicius contrasts Aristotle's account of earth, water, air, and fire in *GC* 2.1-3, with the geometric treatment of them in section 35 of *On the Nature of the Kosmos and the Soul* of Timaeus Locrus (Marg (1972)); cf. Plato's *Timaeus* 53C-56C.

351. See the beginning of the previous chapter, 293a17-b32.

352. i.e. *modus tollens*, the second *anapodeiktos* argument of the Stoics; see e.g. Kneale and Kneale (1962), pp. 162-3. See also the note on 466,32 in the commentary on ch. 9.

353. The star Regulus. In his star catalogue (Heiberg (1898-), vol. 1.2, 98-9, line 6) Ptolemy gives the longitude of Regulus as 2.5 degrees into Leo.

354. cf. 532,35-533,2 with the note.

355. Simplicius says that the two are the same 'in substratum' (*hupokeimenôi*) to indicate that the notion of being the centre of the universe is not the same as the notion of being the centre of the earth.

356. On the question of the category to which angles belong see Heath (1926), vol. 1, pp. 177-8, 252.

357. *El.* 1, def. 10. Heiberg prints *poiêsêi* with A, whereas Euclid, F, and Karsten have *poiêi*.

358. The meaning of this expression (or at least its reference) becomes clear in the geometric argument at 539,5-17. The plane is the surface of the earth. The notion of right angle is extended to include angles like FAC in the figure of that argument.

359. Simplicius has *pros orthas gônias akhthêi eutheia*; in *El.* Heiberg prints *pros orthas [gônias] eutheia grammê akhthêi* (*gônias* being omitted in his preferred manuscript P of the *Elements*).

360. On the equality of the angles contained by a diameter and circumference of a semicircle see Heath (1949), pp. 23-4. Euclid (*El.* 3, def. 7) defines the angle of a segment of a circle, but offers no proofs concerning their equality; presumably the equality of the angles of similar segments of a circle would be proved by superposition; cf. Alexander, *in An. Pr.* (*CAG*, vol. 2.1), 268,15-16.

361. *keratoeideis*, the standard term for the angles made by a convex circumference and a straight line; cf. Proclus, *in Euc.* (Friedlein (1873)), 127,14.

362. This figure is found only in MS A.

363. In *Suntaxis* 1.7 (Heiberg (1898-), vol. 1.1, 21,14-22,11). Ptolemy simply asserts that weights fall in a direction perpendicular to the plane tangent to the earth at the point of contact, and says that this means they would fall to the centre of the earth if they weren't stopped.

364. Heiberg inserts a *hai* from Euclid which is omitted by Simplicius.

365. That is, the diameter of the earth is less than that of the sublunary sphere.

366. It is clear from 540,18 that Simplicius takes this argument to be directed against both the alleged Platonic position that the earth rotates around the central axis and a position according to which it moves in a straight line, but not against the Pythagorean position that the earth is a planet.

367. The lemma omits an *an eiê* of our MSS of Aristotle; it is included by Karsten. It is also omitted by Simplicius at 541,6.

368. The text, which Heiberg prints as *kata tên <kata> diametron autês aei pros ton hêlion stasin apoteleisthai* is obscure, but the sense is assured by the parallel passage in Ptolemy's *Almagest*: *pros tên kata diametron têi hêliôi stasin apoteleisthai* (Heiberg (1898-), vol. 1.1, 19,24-5).

369. For an account of Heraclides' astronomical views see, e.g., Heath (1913), part 1, ch. 18. Most people accept Heath's view that Heraclides assigned to the earth a west-east rotation about the poles of the equator to replace the east-west rotation of the fixed sphere. Simplicius' criticism of this position makes the apparently false assumption that Heraclides denied an independent west-east motion to the sun and other planets.

370. Simplicius virtually quotes the *Almagest* (Heiberg (1898-), vol. 1.1, 25,6-12).

371. I have translated Heiberg's text of Simplicius' citation of 297a34-b1 at 544,13-14: *êremêsei ge kai nun kai mê to meson ekhousa, hêi pephuke kineisthai.* Moraux, following the MSS of Aristotle, prints the words *kai nun* after *kineisthai*, giving the sense 'if it did, it could also be stationary, not occupying the centre to which it is now its nature to move'.

372. I have translated Simplicius' *sumbainei to eirêmenon* (545,11-12) rather than the more difficult *eirêtai to sumbainon* of our MSS of Aristotle.

373. For some textual issues here see 545,20-22 with the note.

374. Heiberg's lemma does not have the *te* of Aristotle (and Karsten), which connects with the *kai* beginning the next lemma.

375. On Alexander's view heavier things push less heavy things away from the centre until a sphere is formed. Simplicius objects that this would mean that the less heavy things were being forced away from the centre, and proposes that the heavier things 'push' the less heavy ones towards the centre and the whole mass 'strives' to get as close to the centre as possible.

376. On this characterisation of the centre of gravity see Heath (1921), pp. 250-1; for Archimedes' works on centres of gravity see his *On the Equilibrium of Planes* (Heiberg (1910-15), vol. 2, 124-212; see also Pappus *Col.* (Hultsch (1876-8)), 1022-46.

377. The difficulty on which Simplicius is now commenting is not easy to interpret. We may state it by supposing that the centre of the earth E is C_e and the centre of the universe is C_u. Now these two centres coincide, and the question is what would happen if E were to be added to, to become E′ with a new centre C_e'. Simplicius takes it that at 297b4-7 Aristotle expresses the view that in this situation E′ would reform into a sphere with centre C_u. But for now Aristotle considers two alternatives; (i) E′ does not 'rest at the centre'; (ii) E′ rests in its new position. For Simplicius (ii) is impossible because if E′ rests, then E would rest if its centre were C_e'. But he does not indicate what, if anything, is wrong with (i), and so does not tell us what he takes Aristotle's first alternative to be.

378. That is, the difficulty does not depend on the amount of weight added on one side of the earth.

379. A comma is needed after *meson*, as in Karsten.

380. Just above at 545,14-15 Simplicius cited these words with *mekhri toutou pheresthai heôs* (move up to the point where), the text of our MSS of Aristotle. An alternative text, apparently rejected by Alexander, is *meta toutou pheresthai heôs* ('move with this until'). Heiberg follows A, in which the alternative accepted by Alexander is *mekhri toude pheresthai heôs* (with the same meaning as *mekhri toutou pheresthai heôs*); F and Karsten have the *toutou* of our MSS of Aristotle. For other passages in the *De Caelo* where, according to Simplicius, Alexander reported divergences among the MSS, see Moraux, p. clxii, n. 1.

381. Heiberg prints *barê* with A and F; Karsten and Moraux print *barea*. Simplicius consistently uses *barê* in the discussion of the lemma.

382. For the lemma Heiberg prints the *mê* of A rather than the *mê ho* of F and Karsten, which is in all the MSS of Aristotle and in Simplicius' paraphrase/citation at 546,13.

383. On the weakness of the argument here see Neugebauer (1975), pp. 1093-4.

384. cf. Theon of Smyrna (Hiller (1878)), 195,5-197,7 and Cleomedes (Todd (1990)), 2.2.19-30.

385. Reading *tinos ginetai* with F and Karsten rather than the *prosginetai* of A printed by Heiberg.

386. That is, Thebes (modern Luxor), one of several Egyptian cities which the Greeks called Diospolis; see PW, vol. 5, cols 144-5.

387. Cadiz.

388. *ton te peri ta Gadeira* [viz. *topon*] *kai tas Hêrakleious stêlas, on Hêrakleian ekalese.* For Aristotle Moraux prints *ton peri tas Hêrakleias stêlas topon*, and Longo (1961) prints what he supposes Simplicius to have read, namely *ton peri tên Hêrakleian topon.*

389. i.e. 400,000.

390. On the determination of the earth's magnitude see below 549,1-10, with the note on 549,10.

391. This figure for the relative volumes of sun and earth is given in 5,16 of Ptolemy's *Almagest* (Heiberg (1898-), vol. 1.2, 427,8-9); it is also in Proclus, *Hypotyposis* (Manitius (1909)), 133,22-4.

392. Heraclitus apparently said that the sun is a foot in breadth (DK22B3). Aristotle knows that it is not, but he twice says it looks that way (*DA* 3.3, 428b2-4; *Somn.* 2, 460b16-20).

393. cf. 512,9-20.

394. On these two instruments, which are very similar, see Rome (1927).

395. The diopter in question is presumably like the one described by Heron in chs 1 to 5 of the *Dioptra* (Schöne (1903), 188,1-204,24). He describes its use for measuring distances between stars in ch. 32 (287,34-288,19).

396. Reading *tous topous* with F and Karsten rather than the *topous* of A, which is printed by Heiberg.

397. Heron describes an odometer in ch. 24 of the *Dioptra* (Schöne (1903), 292,16-302,2).

398. i.e. 180,000. The fixed figures in this description are the 180,000 stades for the circumference of the earth and the equivalent $1^\circ = 500$ stades (cf. Ptolemy, *Geography* (Nobbe (1843)), 1,7,1; 1,11,2; 7,5,12 (Berggren and Jones (2000), pp. 64, 71, 110), and *Planetary Hypotheses* (Goldstein (1967), section 4)). The length of the earth's equator is, in fact, slightly less than 25,000 miles. Assuming that the earth is a sphere with that circumference Ptolemy's figure for the earth's circumference would be incredibly accurate if he used a stade of 733 feet. But we do not know the length of Ptolemy's stade. The 'method' described by Simplicius is certainly a mathematical exercise, not a piece of observational geography. For the method to work the two stars must be on the same celestial longitude, so that the places at which the stars are at the zenith would then be on the same meridian (cf. Ptolemy's description of the method of his predecessors for determining the circumference of the earth at 1,3,1 of the *Geography* (Berggren and Jones (2000), p. 61); on the determination of the meridian of a place see Neugebauer (1975), pp. 841-2). The accuracy of the odometer measurement of the earthly distance (if 'correct', *c.* 70 miles) between the two places would depend on the terrain and the ability of its user to move on a straight line between the two places. The measurement of a distance of 1° between the two stars and the determination of the places for which the stars are at the zenith could not be made with any reasonable accuracy (and an error of one nth of a degree in the former means a one nth error in the estimate of the earth's circumference).

399. In proposition 3 of *The Measurement of a Circle* (Heiberg (1910-15), vol. 1, 236,7-242,21) Archimedes establishes rigorously that the value in question lies between $3^{10}/_{71}$ and $3^{1}/_{7}$. See also the note on 414,4 in the commentary on ch. 4.

400. i.e. 57,273, a slight rounding off.

401. As proposition 1 of *The Measurement of a Circle* (Heiberg (1910-15), vol. 1, 232,1-234,17) Archimedes proves the equivalent, namely that the area of a circle is equal to that of a right-angled triangle with legs equal to the radius and the circumference of the circle.

402. A double myriad is 100 million (= $10{,}000^2$). So the answer calculated is 2,577,285,000.

403. Archimedes, *On the Sphere and Cylinder*, book 1, prop. 33 (Heiberg (1910-15), vol. 1, 120,14-124,13).

404. I have translated the text printed by Heiberg, including the inserted 'simple'. This represents the correct calculation, i.e. 10,309,140,000. All the texts reported by Heiberg (A, F, b, and Karsten) give an answer of 10,307,140,000, i.e. they have *heptakosiôn* instead of the *hennakosiôn* printed by Heiberg.

405. In the next calculation Simplicius applies a part of the porism to proposition 34 of book 1 of Archimedes' *On the Sphere and Cylinder* (Heiberg (1910-15), vol. 1, 130,4-132,3): the volume of a sphere is two-thirds that of a right cylinder with base equal to a greatest circle of the sphere and height equal to its diameter.

406. A triple myriad is a 'quadrillion' (= $10{,}000^3$).

407. i.e. 147,608,843,805,000.

408. At this point Simplicius' calculations go awry. He ends up with 98,406,364,469,503, where he should have 98,405,895,870,000.

409. See the note on 549,4.

410. Reading the *khthamalôtera* of A rather than *khthamalôtata* of F and Karsten printed by Heiberg. With that change this sentence is a virtual quotation of Theon of Alexandria (Rome (1936), 394,17-395,2). The two passages are printed as fr. II A, 2 in Berger (1880). There is a parallel passage without the number in Theon of Smyrna (Hiller (1878), 124,19-22). On Eratosthenes see *DSB*, vol. 4, pp. 388-93.

411. Since Simplicius goes on to say that he doesn't know what unit of measurement Aristotle is using, it is not clear why he chooses to say Aristotle is talking about surface area rather than length of circumference.

Appendix

[Alexander][1] on *Metaphysics* 12.8, 1073b17-1074a15

1073b17-1074a15 Eudoxus postulated that the motion of[2] the sun and moon was in three spheres; [of these the first was that of the fixed stars, the second <rotated> on the circle through the middle of the signs of the zodiac, the third on an inclined circle in the breadth of the zodiac (the circle on which the moon moves is inclined at a greater depth than that on which the sun moves). And he postulated that the motion of each of the planets was in four spheres; and of these, the first and second were the same as those <for the sun and moon>; for the sphere of the fixed stars moves all the spheres, and the sphere positioned under this and having its motion in the circle through the middle of the signs of the zodiac is common to all; the poles of the third sphere were in every case in the circle through the middle of the signs of the zodiac, and the motion of the fourth was on a circle oblique to the middle of this circle; and the poles of the third sphere are different for each except that those of Venus and Mercury are the same.

1073b32 Callippus made the position of the spheres, i.e. the order of their distances,[3] the same as Eudoxus, and he assigned the same number <of spheres> to Jupiter and Saturn as Eudoxus, but he thought that two spheres had to be added to the sun and moon, if one was going to explain the phenomena, and one more to each of the other planets.

1073b38 But if all the spheres added together are going to explain the phenomena, it is necessary that there be for each of the planets other spheres, one fewer <than the number of spheres associated with the planet>, which are counteractive and always restore the first sphere of the star beneath it in order to the same position; for only in this way can they all produce the motion of the planets. So since[4] there are eight spheres in which some of the planets move and twenty-five in which others do, and of these only those in which the low star moves do not need to be counteracted,[5] there will be six spheres counteracting the spheres of the first two and sixteen for the next four. The number of all of the spheres, those carrying the stars and those counteracting them, will be fifty-five. And if one were not to add for the moon and sun the motions which we mentioned, they will make forty-seven in all.[6] <So let the number of the spheres be this great.>[7]]

The views of Eudoxus and Callippus and later Aristotle on the number of spheres, how they move, and why, have been stated with great care in the second book of <my> *De Caelo*[8] in the exegesis of the text extending from 'and furthermore the other motions have a single body because' to 'the power of any 703,1 finite body is related to a finite body'.[9] And that was essentially an exegesis of the present passage extending from 'Eudoxus postulated that the motion of the sun and moon was in three spheres' to 'so let the number of spheres be this great'. So, let the person who wishes provide himself with clarification on this subject from there. But since some of the passages have not been clarified there, it is worthwhile to investigate them here.

The first of these passages is 'of these the first was that of the fixed stars'. Aristotle does not mean by this – as someone might think – that the fixed sphere is first (although it is, in fact, first); rather what he is saying is that, according to Eudoxus, of the three spheres carrying the sun, the first, which both contains the other two and has the same motion as the fixed sphere (for they said it moves from east to west and we said why in <my> *De Caelo*), was considered by Eudoxus to be fixed relative to the other two which it contained. And he didn't call just the first sphere of the sun fixed but <he called> also the first sphere of Saturn and of Jupiter, and the first and greater spheres of each of the other stars fixed, the first sphere being one fixed sphere in which are the mass of stars filling up the zodiac, the first sphere for Saturn another, that for Jupiter another, and so on – Theophrastus called these spheres starless.[10]

The words 'the third on an inclined circle in the breadth of the zodiac' mean the circle inclined in the breadth of the signs of the zodiac, as explained in <my> *De Caelo*, the one which the sun is thought to describe with its own central point when it is moved by the sphere in which it is implanted; the tropics seem to vary because when the sun makes its turnings it is not always observed to rise from the same places.

The words 'the circle on which the moon moves is inclined at a greater depth than that on which the sun moves' are equivalent to 'the breadth of the inclination of the circle which the moon is thought to describe with its own central point is greater than the breadth of the circle which the sun is thought to describe with its own central point'; and this was learned from the fact that the moon and the sun in their turnings do not make their risings from the same positions. For let the horizon be AGBCMD,[11] the equator DEB, the inclined circle in the breadth of the zodiac which the sun is thought to describe with its own central point AEC, and the inclined circle which the moon is thought to describe with its central point GEM. And let the sun rise at the time when it makes its summer turning from the point A and rise at the equinox at the point D. It is clear that the greatest breadth of inclination is the arc DA. But let the moon when it makes its turning in the north rise at the point G; it is clear that the greatest breadth of its inclination is DG, and it is greater than the inclination DA, which is the greatest breadth of the inclination of the sun. And this is what is meant by 'the circle in which the moon moves is inclined at a greater angle than that in which the sun moves'.

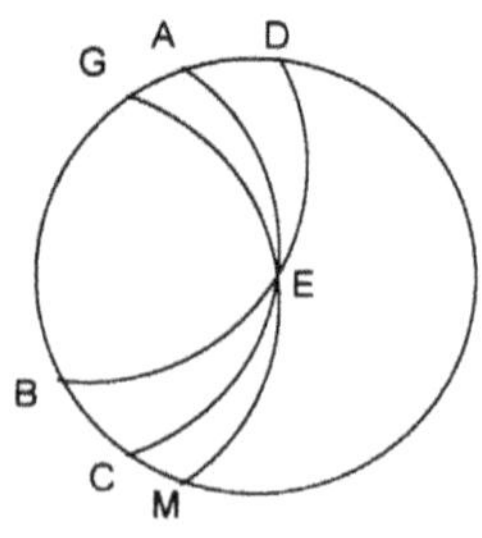

Figure 6 (703,34-704,4)[12]

The words 'the poles of the third sphere were in every case in the circle through the middle of the signs of the zodiac' mean that all the third spheres have their poles in the circles which are described through the signs of the zodiac by the central points of the stars.

(1073b32) The words '<Callippus> made the order of the spheres, i.e. the order of their distances'[13] mean that Callippus said that the kind of spheres which Eudoxus said were first and second were first, and that Callippus said that the sphere of Saturn was as far from the sphere of Mercury as Eudoxus had said.

(1073b38) The words 'that there be for each of the planets other spheres, one fewer <than the number of spheres associated with the planet>, which are counteractive' mean that since there are four spheres moving Saturn, there are three counteractive spheres, since three is less than four by one; and again since according to Callippus there are five spheres moving Mars four counteractive spheres are added to them, since four are one less than five. What purpose the addition of the counteractive spheres served for Aristotle has been explained in <my> *De Caelo*.

The words 'always restore the first sphere of the star beneath it in order to the same position' will be clear if certain circles are drawn. Let there be three circles AB, DE, FG. Let there be a star, e.g. Jupiter, in FG, and let it be the point K.[14] Let A be some point, and D and F and B, E, G. They are not to be stars but are to be conceived just as points, and let A, D, and F lie on the same straight line; and similarly for B, E, G. And let the sphere AB move towards the point B so that A comes to be at point B. And let the sphere DE and the sphere FG move at the same speed in the direction of D, F so that E, G reach D, F. So since ED and GF move in the same direction at the same speed, it is clear that the motion of the sphere GF will be double that of the sphere ED. For the motion of the sphere ED imparts as much motion to the sphere GF as GF has, since they are moving in the same direction. Consequently in the time in which E moves to D, G, moving through the whole circle GF, again reaches G[15] unless the sphere AB moves in an opposite direction to DE and drags it towards B and prevents it from imparting to the sphere FG another motion as great as it has. So AB drags DE towards B[16] and impedes it, and DE impedes FG and in this case FG will always keep the same position in relation to the position of AB.[17] Consequently when G reaches F and F is now where G is, then A will come to be where B is now, and B where A is and the points A, F and B, G will always be on the same 705,1
straight line. And when A rises, so does F and when B sets so does G, and they will never rise or set before A, F. And if AB and GF move in the same direction the same thing will happen.[18]

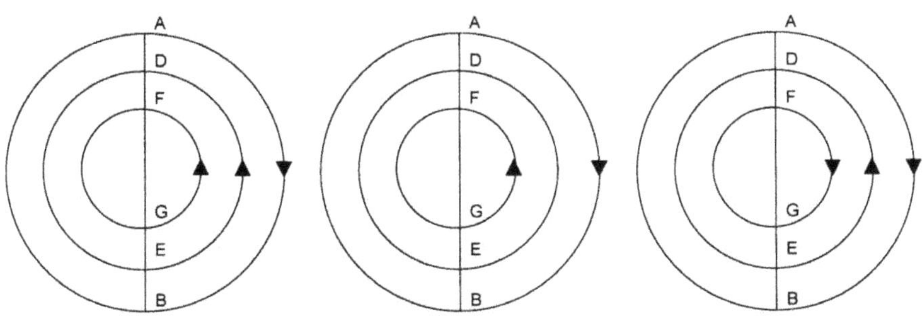

Figure 7 (704,23-35; 704,35-705,4; 704,4-5)[19]

But let the sphere be conceived as counteractive (this then is what is meant by 'always restore the first sphere of the star beneath it in order to the same position'). What is said would be the following. Let the counteractive spheres be conceived to move in such a way that they make the first and outermost of the spheres moving Jupiter always keep the same position in relation to the sphere in which the star is fastened[20] and never let points taken in the first sphere precede the points taken in the sphere in which the star is fastened, but let them always rise and set together.

The words 'so[21] there are eight spheres in which some of the planets move and twenty-five in which others do' are equivalent to 'so according to Callippus there are eight spheres for the sun and moon and twenty-five for the others, Saturn, Jupiter, and the rest'. For what was said just before, 'he thought that two spheres had to be added to the sun and moon' are equivalent to 'one to each'. For since Eudoxus postulated six for the sun and the moon, and he says Callippus postulated eight, it is clear that he added one for each of them. And similarly he added one to the remaining spheres of the five stars. For Eudoxus said there were four spheres for each of the remaining five stars, and <Callippus> made them five.[22]

The words 'and of these only those in which the low star moves do not need to be counteracted'[23] mean 'it is only not necessary for the spheres which move the moon to have counteracting spheres' since the moon is the lowest of all the stars.

The words 'there will be six spheres counteracting the spheres of the first two' mean 'there will be six spheres counteracting Saturn and Jupiter'. For, since there are four spheres moving Saturn and four moving Jupiter and it is necessary that for each star there be one less counteractive sphere, there will be six counteractive spheres for Saturn and Jupiter, and since there are five spheres moving each of the remaining four, that is Mars, Venus, Mercury, and the sun, there will be four counteractive spheres for each.

As a result the total number of counteractive spheres is two times three for Saturn and Jupiter and four times four for Mars, Venus, Mercury, and the sun. So the total is twenty-two. And the carrying spheres were eight for Saturn and Jupiter and twenty-five for the remaining five. So if these thirty-three spheres are added to the twenty-two counteractive spheres, the total will be fifty-five. For, as has been said, only the spheres which move the moon do not have to be counteracted, since the moon is last. And so Aristotle has said that only the spheres in which the planet ordered low is carried do not need to be counteracted.

The additional words 'if one were not to add[24] for the moon and sun the motions which we mentioned, the total number of spheres will be forty-seven'[25] create consternation. For if we subtract the two spheres of the sun and moon which Callippus added (and obviously the two additional counteractive spheres for the sun – for if we subtract the spheres added by Callippus we must also subtract the ones intended to counteract them), six spheres will be subtracted, two carrying the sun, two counteracting them, in addition to the two added by Callippus for the moon. But when these are subtracted from the fifty-five it will not result that there remain forty-seven in all, but forty-nine.[26]

Or perhaps he has forgotten that he subtracted only two, not four, spheres for the moon, unless one should say that he subtracted the four counteractive spheres which he added for the sun and both <pairs> which Callippus added, so that eight spheres were subtracted from the fifty-five, leaving forty-seven as a remainder, and in this way the number which Aristotle gives results.[27] Or, as Sosigenes says with understanding, it is better to say of the number that it is an error produced by the scribes than to make the seven spheres themselves also eight.[28]

Notes

1. The identity of the author of the commentary on books E-N of the *Metaphysics* which is printed as part of a commentary on the whole of the *Metaphysics* and ascribed on the title page to Alexander of Aphrodisias in vol. 1 of *CAG* is a matter of dispute, although there is general agreement that the author of the latter part of the commentary is not Alexander. Tarán (1987) summarises the history of the dispute and argues that the commentary is a forgery produced no later than the mid-fifth century. Hadot (1987a), p. 229, n. 12, and pp. 242-5, defends the more common view that the commentary is the work of Michael of Ephesus, a Byzantine commentator on Aristotle (twelfth century).

2. The lemma omits the words 'each of' found in our MSS of Aristotle.

3. These words (*tout' esti tôn apostêmatôn tên taxin*) are bracketed by Ross (1953) and Jaeger (1957), following a suggestion of Christ (1866). For [Alexander]'s text see 704,9.

4. On [Alexander]'s text here see the note on 705,13.

5. For [Alexander]'s text of this clause see the note on 705,24.

6. On [Alexander]'s text here see the two notes on 705,40-1.

7. See the two notes on 506,4 of Simplicius' commentary. This last sentence actually begins the next lemma in the commentary, but [Alexander] refers to it at 703,6-7.

8. Since the authorship of the present passage is disputed (see the first note in this appendix), it is not possible to know what commentary is being referred to. Moraux ((2001), pp. 224-5) argues that the reference here is to Alexander's commentary on the *De Caelo* (because, as he thinks, the *Metaphysics* commentary we have incorporates much material from Alexander's original commentary). Moraux further infers that the extensive material on planetary theory in Simplicius' commentary, which Simplicius refers to Sosigenes, a teacher of Alexander, was also taken from Alexander. It is perhaps worth mentioning that Simplicius himself wrote a commentary on the *Metaphysics*; see Hadot (1987a).

9. *Cael.* 293a4-11. See the first note on the lemma at 491,12.

10. 703,17-23 are Theophrastus text 165A in Fortenbaugh et al. (1992).

11. The text reads 'AGBMCD', but if that order is used in the diagram the moon would never be south of the sun's path in the zodiac.

12. I have supplied the figures in this appendix.

13. [Alexander] writes *tên autên etitheto taxin, toutesti tôn apostêmatôn*, Ross (1953) and Jaeger (1957) print *tên men thesin tôn sphairôn tên autên etitheto Eudoxôi,* [*tout' esti tôn apostêmatôn tên taxin*].

14. The point K plays no role in the subsequent argument, and I have not included it in the figure.

15. This remark requires that ADFGEB be a diameter of the circle AB, as in Figure 7.

16. i.e. 'towards E'.

17. This will not happen unless DE has only the motion imparted to it by AB, as in Figure 7 (centre).

18. In this case AB and GF cannot keep the same relative position unless DE counteracts the motion of FG, as in Figure 7 (right).

19. In these figures arrows represent self-motion. Containing circles impart whatever motion they have to contained ones. An argument similar to the one given here, due to Sosigenes, is to be found in Simplicius at 499,17-500,14.

20. It is tempting to emend the text here and in the next line. The point of the counteractive spheres for Saturn is to make the first and outermost of the spheres moving Jupiter always keep the same position in relation to the sphere of the fixed stars; so perhaps one should read 'the sphere in which the stars are fastened' (*endedentai hoi asteres* in place of *endedetai ho astêr*).

21. [Alexander] has *hôste* where our texts of Aristotle have *epei oun*.

22. What is said in this paragraph is a misunderstanding; it is corrected starting at 705,25. The meaning should be 'there are eight spheres for Jupiter and Saturn, and twenty-five for the other five planets'.

23. [Alexander] has *anelikhthênai ou dei, en hais to katô pheretai*, where our texts of Aristotle have *ou dei anelikhthênai en hais to katôtatô tetagmenon pheretai*. But at 705,38-9 [Alexander] has *ou dei anelikhthênai en hais pheretai to katô tetagmenon astron*.

24. Aristotle has *prostitheiê* where [Alexander] (and Simplicius at 503,11) has *prostheiê*.

25. [Alexander] (and Simplicius at 503,12) has *hepta kai tettarakonta esontai pasai* where Aristotle has *hai pasai sphairai esontai hepta te kai tessarakonta*.

26. This paragraph is essentially identical with Simplicius 503,10-20.

27. This sentence is very much the same as Simplicius 503,21-26.

28. This sentence is a somewhat garbled version of Simplicius 503,35-504,1, on which see the second note.

Bibliography

Allan, D. J. (ed.) (1934), *Aristotelis De Caelo*, Oxford: Clarendon Press.

Allan, D. J.(1950), 'Mediaeval Versions of Aristotle "De Caelo" ' and of the Commentary of Simplicius', *Mediaeval and Renaissance Studies* 2, 82-120.

Arnim, Hans von (ed.) (1903-1924), *Stoicorum Veterum Fragmenta*, 4 vols, Leipzig: Teubner.

Aujac, Germaine (ed. and trans.) (1975), *Geminos, Introductions aux Phénomènes*, Paris: Belles Lettres.

Aujac, Germaine (ed. and trans.) (1979), *Autolycus de Pitane, La Sphère en Mouvement, Levers et Couchers Héliaques, Testimonia*, Paris: Belles Lettres.

Baltes, Matthias (1972), *Timaios Lokros: Über die Natur des Kosmos und der Seele*, Leiden: E.J. Brill.

Beaujeu, Jean (ed. and trans.) (1950), *Pline l'Ancien, Histoire Naturelle*, Paris: Belles Lettres.

Berger, Hugo (1880), *Die Geographischen Fragmente des Eratosthenes*, Lepizig: Teubner.

Berggren, J. Lennart and Jones, Alexander (trans.) (2000), *Ptolemy's Geography*, Princeton and Oxford: Princeton University Press.

Bergk, Theodor (1842), Review of Karsten (1838) in *Zeitschrift für die Altertumswissenschaft* 9, cols 1001-11. Reprinted in Bergk (1886), pp. 45-59.

Bergk, Theodor (1883), *Fünf Abhandlungen zur Geschichte der Griechischen Philosophie und Astronomie*, Leipzig: Fues's Verlag.

Bergk, Theodor (1886), *Kleine Philologische Schriften*, vol. 2, Halle an der Saale: Waisenhaus.

Bidez, J. (1935), 'Les couleurs des planètes dans le mythe d'Er au livre X de la *République* de Platon', *Académie royale de Belgique: Bulletins de la Classe des Lettres et des Sciences Morales et Politiques*, 5e série, 20, 257-77.

Böckh, August (1863), *Ueber die Vierjährigen Sonnenkreise der Alten*, Berlin: Georg Reimer.

Boll, F. (1917-19), 'Kronos-Helios', *Archiv für Religionswissenschaft* 19, 342-6.

Bossier, Fernand (1987), 'Traductions Latines et influences du commentaire *in De Caelo* en occident (XIIIe-XIVe s.)', in Hadot (1987), pp. 289-325.

Bousset, D.W. (1915), *Jüdisch-Christlicher Schulbetrieb in Alexandria und Rom*, Göttingen: Vandenhoeck & Ruprecht.

Burkert, Walter (1972), *Lore and Science in Ancient Pythagoreanism* (trans. Edwin L. Minar Jr), Cambridge, MA: Harvard University Press.

Cherniss, Harold (1959-60), 'Plato (1950-1957)', *Lustrum* 4 (1959) and 5 (1960).

Christ, W. (ed.) (1886), *Aristotelis Metaphysica*, Leipzig: Teubner.

Cornford, Francis Macdonald (1937), *Plato's Cosmology*, London: K. Paul, Trench, Trubner, and New York: Harcourt, Brace.

Daremberg, Ch., and Saglio, Edm. (eds) (1875), *Dictionnaire des Antiquités*

Grecques et Romaines d'après les Textes et les Monuments, 2nd edn, 5 vols, Paris: Hachette.

Deubner, Ludovicus (ed.) (1975), *Iamblichi De Vita Pythagorica Liber* (with additions and corrections by Udalricus Klein), Stuttgart: Teubner.

Dicks, D.R. (1970), *Early Greek Astronomy to Aristotle*, Ithaca, NY: Cornell University Press.

Diehl, Ernestus (ed.) (1903-6), *Procli Diadochi in Platonis Timaeum Commentaria*, 3 vols, Leipzig: Teubner.

Dillon, John (1987), 'Iamblichus of Chalcis (*c*. 240-325 AD)', in Wolfgang Haase (ed.), *Aufstieg und Niedergang der Römischen Welt*, II 36.2, Berlin and New York: Walter De Gruyter, pp. 862-909.

Dillon, John (1996), *The Middle Platonists*, revised edn, Ithaca NY: Cornell University Press, and London: Duckworth.

Düring, Ingemar (ed.) (1930), *Die Harmonielehre des Klaudios Ptolemaios* (Göteborgs Högskolas Årsskrift 36), Göteborg: Elanders Boktryckeri Aktiebolag.

Endress, Gerhard, and Kruk, Remke (eds) (1997), *The Ancient Tradition in Christian and Islamic Hellenism*, Leiden: Research School CNWS, School of Asian, African, and Amerindian Studies.

Fontaine, Jacques (ed. and trans.) (1960), *Isidore de Seville, Traité de la Nature*, Bordeaux: Féret et Fils.

Fortenbaugh, W.W. et al. (ed. and trans.) (1992), *Theophrastus of Eresus: Sources for his Life, Writings, Thought, and Influence*, Leiden: E.J. Brill.

Fränkel, Hermann (ed.) (1961), *Apollonii Rhodii Argonautica*, Oxford: Clarendon Press.

Friedlein, Godofredus (ed.) (1873), *Procli Diadochi in Primum Euclidis Elementorum Librum Commentarii*, Leipzig: Teubner.

Görgemanns, Herwig (1970), *Untersuchungen zu Plutarchs Dialog De facie in orbe lunae* (Bibliothek der klassischen Altertumswissenschaften n.f., 2. Reihe, Bd. 33), Heidelberg: C. Winter.

Goldstein, Bernard R. (1967), *The Arabic Version of Ptolemy's Planetary Hypotheses* (Transactions of the American Philosophical Society n.s. 57.4), Philadelphia: The American Philosophical Society.

Goulet, Richard (ed.) (1989-), *Dictionnaire des Philosophes Antiques*, Paris: Éditions du Centre National de la Recherche Scientifique.

Hadot, Ilsetraut (ed.) (1987), *Simplicius, sa Vie, son Oeuvre, sa Survie*, Berlin and New York: Walter de Gruyter.

Hadot, Ilsetraut (1987a), 'Recherches sur les fragments du commentaire de Simplicius sur la *Métaphysique* d'Aristote', in Hadot (1987), pp. 225-45.

Hadot, Ilsetraut (1990), 'The Life and Work of Simplicius in Greek and Arabic Sources', in Sorabji (1990), pp. 275-303.

Hall, J.J. (1971), ' "Planets" in Simplicius *de caelo* 471.1ff.', *Journal of Hellenic Studies* 91, 138-9.

Hartner, Willy (1964), 'Mediaeval views on cosmic dimensions and Ptolemy's Kitab Al-Manshurat', in *Mélanges Alexandre Koyré*, pp. 254-82.

Heath, Thomas (1913), *Aristarchus of Samos*, Oxford: Clarendon Press.

Heath, Sir Thomas (1921), *A History of Greek Mathematics*, 2 vols, Oxford: Clarendon Press.

Heath, Thomas (trans.) (1926), *The Thirteen Books of Euclid's Elements*, 2nd edn, 3 vols, Cambridge: Cambridge University Press.

Heath, Sir Thomas (1949), *Mathematics in Aristotle*, Oxford: Clarendon Press.

Heiberg, J.L. (1892), 'Handschriftliches zum Commentar des Simplicius zu

Aristoteles *de caelo*', *Sitzungsberichte der königlich preussischen Akademie der Wissenschaften zu Berlin*, Berlin: Georg Reimer, pp. 59-66.

Heiberg, J.L. (ed.) (1898-), *Claudii Ptolemaei Opera Quae Exstant Omnia*, 3 vols, Leipzig: Teubner.

Heiberg, J.L. (ed.) (1910-15), *Archimedis Opera Omnia*, 2nd edn, 3 vols, Leipzig: Teubner.

Heiberg, J.L. (ed. and trans.) (1927), *Theodosius Tripolites Sphaerica* (*Abhandlungen der Gesellschaft der Wissenschaften zu Göttingen* Philologische-historische Klasse, Neue Folge 19.3), Berlin: Weidmann.

Heiberg, J.L., and Menge, Hermann (ed. and trans.) (1883-1916), *Euclidis Opera Omnia*, Leipzig: Teubner.

Hiller, Eduardus (ed.) (1878), *Theonis Smyrnaei Philosophi Platonici Expositio Rerum Mathematicarum ad Legendum Platonem Utilium*, Leipzig: Teubner.

Hoffmann, Philippe (1981), *Recherches sur la Tradition Manuscrite du Commentaire de Simplicius au De Caelo d'Aristote*, Thèse présentée à l'Université de PARIS IV pour le Doctorat de Troisième Cycle.

Huffman, Carl A. (1993), *Philolaus of Croton*, Cambridge: Cambridge University Press.

Hultsch, Fridericus (ed.) (1876-8), *Pappi Alexandrini Collectionis Quae Supersunt*, 3 vols, Berlin: Weidmann.

Ideler, Ludwig (1830), 'Über Eudoxus', *Abhandlungen der königlichen Akademie der Wissenschaften zu Berlin*, Historisch-philologische Klasse, pp. 49-88 (published in 1832).

Jaeger, W. (ed.) (1957), *Aristotelis Metaphysica*, Oxford: Clarendon Press.

Jenkins, Ian (1986), *Greek and Roman Life*, Cambridge, MA: Harvard University Press.

Kahn, Charles H. (1960), *Anaximander and the Origins of Greek Cosmology*, New York and London: Columbia University Press.

Karsten, Simon (ed. and trans.) (1838), *Empedoclis Agrigentini Carminum Reliquae*, Amsterdam: Johannes Müller.

Kidd, Douglas (ed. and trans.) (1997), Aratus, *Phaenomena* (Cambridge Classical Texts and Commentaries 34), Cambridge: Cambridge University Press.

Kneale, William, and Kneale, Martha (1962), *The Development of Logic*, Oxford: Clarendon Press.

Lameer, Joep (1997), 'From Alexandria to Baghdad: Reflections on the Genesis of a Problematical Tradition', in Endress and Kruk (1997), pp. 181-91.

Lasserre, François (ed. and trans.) (1966), *Die Fragmente des Eudoxos von Knidos*, Berlin: Walter de Gruyter.

Last, Hugh (1924), 'Empedocles and his Klepsydra Again', *Classical Quarterly* 18, 169-73.

Longo, Oddone (ed. and trans.) (1961), *Aristotele, De Caelo*, Firenze: Sansoni.

Manitius, Carolus (ed. and trans.) (1909), *Procli Diadochi Hypotyposis Astronomicarum Positionum*, Stuttgart: Teubner.

Manitius, Carolus (ed. and trans.) (1894), *Hipparchi in Arati et Eudoxi Phaenomena Commentariorum*, Leipzig: Teubner.

Marg, Walter (ed. and trans.) (1972), *Timaeus Locrus: De Natura Mundi et Animae*, Leiden: E.J. Brill.

Mélanges Alexandre Koyré, Paris: Hermann, 1964.

Mendell, Henry (1998), 'Reflections on Eudoxus, Callippus, and their Curves: Hippopedes and Callippopedes', *Centaurus* 40, 177-275.

Mendell, Henry (2000), 'The trouble with Eudoxus', in Suppes, Moravcsik, and Mendell (2000), pp. 59-138.

Mioni, Elpidio (1985), *Bibliothecae Divi Marci Venetiarum Codices Graeci Manuscripti*, vol. 1, Rome: Istituto Poligrafico dello Stato.
Moraux, Paul (1954), 'Notes sur la tradition indirecte du "De caelo" d'Aristote', *Hermes* 82, 145-82.
Moraux, Paul (1984), *Der Aristotelismus bei den Griechen*, vol. 2, Berlin and New York: Walter de Gruyter.
Moraux, Paul (2001), *Der Aristotelismus bei den Griechen*, vol. 3, Berlin and New York: Walter de Gruyter.
Neugebauer, Otto (1975), *A History of Ancient Mathematical Astronomy*, 3 vols, Berlin: Springer-Verlag.
Nobbe, Carolus Fridericus Augustus (ed.) (1843), *Claudii Ptolemaei Geographia*, Leipzig: Tauchnitz.
Pedersen, Olaf (1974), *A Survey of the Almagest* (Acta Historica Scientiarum Naturalium et Medicinalium 30), Odense: Odense Universitetsforlag.
Peyron, Amedeo (1810), *Empedoclis et Parmenidis Fragmenta*, Leipzig: I.A.G. Weigl.
Pottier, Edmond, and Reinach, Salomon (1887), *La Necropole de Myrina*, 2 vols, Paris: Ernest Thorin.
Prandi, Luisa (1985), *Callistene. Uno storico tra Aristotele e i re macedoni*, Milan: Jaca Book.
Rome, A. (1927), 'L'Astrolabe et le Météoroscope d'après le commentaire de Pappus sur le 5e livre de l'Almageste', *Annales de la Société Scientifique de Bruxelles*, vol. 47, series A (Sciences Mathématiques), part 2 (Mémoires), 77-102.
Rome, A. (ed.) (1936), *Commentaires de Pappus et de Théon d'Alexandrie sur L'Almageste*, vol. 2 (*Studi e Testi* 72), Vatican City: Biblioteca Apostolica Vaticana.
Rose, Valentinus (1886), *Aristotelis Qui Ferebantur Librorum Fragmenta*, 3rd edn, Lepizig: Teubner.
Ross, W.D. (1953), *Aristotle's Metaphysics: A Revised Text with Introduction and Commentary*, 2 vols, Oxford: Clarendon Press.
Schiaparelli, Giovanni (1874), 'Le sfere omocentriche di Eudosso, di Callippo e di Aristotele', reprinted in Schiaparelli (1925-27), vol. 2, pp. 3-112 (German translation by W. Horn as 'Die homocentrischen Sphären des Eudoxos, des Kallippus und des Aristoteles', *Abhandlungen zur Geschichte der Mathematik* 1 (1877), 101-98).
Schiaparelli, Giovanni (1925-27), *Scritti sulla storia della astronomia antica*, 3 vols, Bologna: Nicola Zanichelli.
Schoch, Karl (1927), *Planeten-tafeln für Jedermann*, Berlin: Linser Verlag.
Schöne, Hermann (ed. and trans.) (1903), *Herons vom Alexandria Vermessungslehre und Dioptra*, Leipzig: Teubner.
Sharples, R.W. (1987), 'Alexander of Aphrodisias: Scholasticsm and Innovation', in Wolfgang Haase (ed.), *Aufstieg und Niedergang der Römischen Welt*, II 36.2, Berlin and New York: Walter De Gruyter, pp. 1176-1243.
Smith, Andrew (ed.) (1993), *Porphyrii Philosophi Fragmenta*, Stuttgart and Leipzig: Teubner.
Snell, Bruno et al. (ed.) (1971-), *Tragicorum Graecorum Fragmenta*, Göttingen: Vandenhoeck & Ruprecht.
Sorabji, Richard (ed.) (1990), *Aristotle Transformed: The Ancient Commentators and Their Influence*, London: Duckworth, and Ithaca, NY: Cornell University Press.
Stein, Heinrich (ed.) (1852), *Empedoclis Agrigentini Fragmenta*, Bonn: A. Marcus.

Stratton, George Malcolm (1917), *Theophrastus and the Greek Psychological Theory before Aristotle*, London: George Allen and Unwin.

Suppes, Patrick, Moravcsik, Julius M., and Mendell, Henry (eds) (2000), *Ancient and Medieval Traditions in the Exact Sciences*, Stanford: CSLI Publications.

Tarán, Leonardo (1987), 'Syrianus and Pseudo-Alexander's commentary on Metaph. E-N', in Wiesner (1987), pp. 215-32.

Tardieu, Michel (1990), *Les Paysages Reliques: routes et haltes syriennes d'Isidore à Simplicius*, Louvain and Paris: Peeters.

Taylor, A.E. (1928), *A Commentary on Plato's Timaeus*, Oxford: Clarendon Press.

Thiel, Helmut van (ed.) (1991), *Homeri Odyssea*, Hildesheim and New York: G. Olms.

Todd, Robert (ed.) (1990), *Cleomedis Caelestia*, Leipzig: Teubner.

Toomer, G. J. (1988), 'Hipparchus and Babylonian Astronomy', in Erle Leichty, Maria de J. Ellis, and Pameli Gerardi (eds), *A Scientific Humanist: Studies in Memory of Abraham Sachs* (Occasional Publications of the Samuel Noah Kramer Fund 9), Philadelphia: no publisher given, pp. 353-62.

Verdenius, W. J. (1969), 'Critical and Exegetical Notes on *De Caelo*', in Ingemar Düring (ed.), *Naturphilosophie bei Aristoteles und Theophrast*, Heidelberg: Lothar Stiehm Verlag.

Vlastos, Gregory (1975), *Plato's Universe*, Seattle: University of Washington Press.

Wartelle, André (1963), *Inventaire des Manuscrits Grecs d'Aristote et de ses Commentateurs*, Paris: Belles Lettres.

Wehrli, Fritz (1955), *Die Schule des Aristoteles*, vol. 8 (Eudemos), Basel: Benno Schwabe & Co.

West, M.L. (ed.) (1966), *Hesiod, Theogony,* Oxford: Clarendon Press.

West, M.L. (ed.) (2000) *Homeri Ilias*, vol. 2, Munich and Leipzig: K.G. Saur.

Wiesner, Jürgen (ed.) (1987), *Aristoteles – Werk und Wirkung*, vol. 2, Berlin and New York: Walter De Gruyter.

Willamowitz-Moellendorff, U. v. (1930), 'Lesefrüchte', *Hermes* 65, 241-58.

Willis, James (ed.) (1983), *Martianus Capella*, Leipzig: Teubner.

Wright, M.R. (1981), *Empedocles: The Extant Fragments*, New Haven and London: Yale University Press.

Yavetz, Ido (1998), 'On the homocentric spheres of Eudoxus', *Archive for History of Exact Sciences* 52, 221-78.

Yavetz, Ido (2001), 'A new role for the hippopede of Euxous', *Archive for History of Exact Sciences* 56, 69-93.

Textual Questions

(a) Textual suggestions

Listed here are places where I have translated a text different from the one printed by Heiberg. For reasons which I hope I have made clear in the Introduction I prefer to call these deviations suggestions rather than emendations. In many cases notes on the lines in the translation provide more information.

471,9 For *metaparabolês* read *parabolês* with D, E, F, and Karsten.

474,9 For *Areos* one should perhaps read *Hêliou.*

474,28 For *ek* read *<apostaseôs> ek.*

480,22 For *tas* read *toiautas tas,* a suggestion of Heiberg.

482,24 For *hêi* read *ê* with A, C, D, E, F, and Karsten.

486,13 For *ep' auto* read *auto* with Karsten.

488,28 For *aplanôn* read *planômenôn* or *planêtôn.*

490,5 For *eirêken* read *eilêptai* with D and F (Karsten has *eilêphen*).

491,23 For *idion* read *idios* with D and Aujac (1979) (Karsten has *idion*).

494,28 For *mesou* read *mesôn* with D,E,F, Karsten, and Aujac (1979).

495,15 For *dusmôn* read *dusmas* with D, E, F, Karsten, and Aujac (1979).

498,2 For *houtôs* read *houtôs kai* with Karsten (D, E, and F have *houtô kai*).

499,19 For *menousês* read *kinoumenês* with F and Karsten; for *perieilousês* read *menousês tês periekhousês* with F and Karsten.

499,22 For *menousa* read *menonta* with F and Karsten.

502,3 For *tên dia tettarôn* read *ton Dia* [*tettarôn*].

502,15 For *deutera* read *hebdoma.*

502,16 For *prosetithei* read *prostithêsi.*

503,33 For *anelittousan* read *anelittousas.*

503,34 For *Metaphusikês* read *Meta ta phusika* with F and Karsten.

504,10 For *diaphoras* read *phoras* with Karsten.

505,27 For *auton* read *autôi* with A (F and Karsten have *auton*).

508,1 For *ekkentrou* read *kentrou.*

508,12 Bracket *perigeiou* (Karsten has a preferable *prosgeiou*).

508,22 For *pros to* read *tôi* with F, Karsten, and Aujac (1979).

513,24 For *prosêkein* read *prosêkei* with F and Karsten.

513,29 For *tên autên* read *tautên tên* with A, F, B, and Karsten.

515,7 For *legei* read *legôn* with F and Karsten.

518,2 For *sêmainousês* read *to strephomenês sêmainousês* with F and Karsten.

518, 3 For *eirêmenou* read *eirêmenon* with F and Karsten.

518,4 For *strephomenên* read *sustrephomenên* with F (Karsten has *strephomenên*).

518,13 For *khrêseis* read *rhêseis* with Karsten.

518,25 Place a period after *ekdekhomenous,* and read *oun* for *an* with A, F, and Karsten.

518,26 For *tautên* read *tên gên* with F and Karsten.
521,14 For *peri tas peri toutou* read *peri tas toutôn* with Karsten (F has *tas peri toutou*).
521,19 For *toutou* read *toutôn* with F and Karsten.
521,24 For *toutou* read *toutôn* with F and Karsten.
524,29 For *eisienai* read *eisienai aera* with F and Karsten.
525,9 For *enseisthentos autois* read *enseisantos auta*.
526,7 For *anêirêke* read *aneirêke* with F, Karsten, and a correction in A.
527,4 For *epizountes* (presumably a misprint) read *tên aitian zêtountes* with F and Karsten.
527,6 For *toutou* read *tou ouranou* with F and Karsten.
533,34 For *anairoumetha* read *airoumetha* with F and Karsten.
535,11 For *dinêseôs* read *dinês* with F and Karsten.
535,23 For the first *to* read *ton* with Karsten.
547,4 For *prosginetai* read *tinos ginetai* with F and Karsten.
549,7 For *topous* read *tous topous* with F and Karsten.
550,2 For *khthamalôtata* read *khthamalôtera* with A (F and Karsten have *khthamalôtata*).

(b) Simplicius' citations of *On the Heavens* 2.10-14

Here I bring together places where the text of a citation by Simplicius of a passage from *De Caelo* 2.10-14 as printed by Heiberg differs from Moraux's text of Aristotle. I also indicate what is printed by Karsten and (where Heiberg provides the information) what appears in F. In general Heiberg's text reproduces A. I have not included a few cases where it seems clear that Simplicius had a text different from ours, but there is no citation, and I have paid no attention to the numerous differences regarding elision (e.g. *de* vs. *d'*) or minor variations in spelling (e.g. *hauton* vs. *heauton* or *teleiotaton* vs. *teleôtaton*).

Moraux	Heiberg	Karsten
292a14 *dê*	481,21 *men dê*	*dê*
292b4 *Tôi d' hôs*	484,5 *tôi de isôs*	*tôi de isôs*
292b6 *de*	484,10 *gar*	*de*
292b6 *aei estin*	484,10 omit	*aei estin*
292b7 *Tôn d'*	484,16 *kai dê kai tôn*	*kai dê kai tôn*
294b33 *tis*	526,14 omit	*tis* (and F)
295a30 *eipeien*	528,5 *eipoi an*	*an eipoi* (and F)
297b1 *mê*	544,13 *nun kai mê*	*nun kai mê*
297b1 *kai nun*	544,14 omit	omit
297b9 *eirêtai to sumbainon*	545,11 *sumbainei to eirêmenon*	*sumbainei to eirêmenon*

(c) Simplicius' citations of other texts

Here I bring together places where the text of a citation by Simplicius of a passage from a work other than *De Caelo* as printed by Heiberg differs from the text of a standard edition of the work. I also indicate what is printed by Karsten and (where Heiberg provides the information) what appears in F. In general Heiberg's text reproduces A. I have paid no attention to the numerous differences regarding elision (e.g. *de* vs. *d'*) or minor variations in spelling (e.g. *hauton* vs. *heauton* or *teleiotaton* vs. *teleôtaton*).

Aristotle, *De Caelo* 2.1-9

Moraux	Heiberg	Karsten
292a19 *autôn monon*	378,14 *monon autôn* 388,23 *monon autôn*	*monon autôn* (also F) *monon autôn*
292a19 *monadôn*	388,23 *monadôn dianooumetha*	*monadôn dianooumetha*
292a20 *pampan, dianooumetha*	388,24 *pampan*	*pampan*

Aristotle, *Metaphysics*

Ross (1953)	Heiberg	Karsten
1073b11 *men*	505,30 *oun men*	*men*
1073b13 *hupolabein*	505,32 *hupolambanein*	*hupolabein*
1073b14 *para*	506,1 omit	*para*
1073b37 *mellei*	497,13 *melloi*	*mellei* (also E)
1073b38 *mian*	497,10 *ana mian*	*mian*
1074a3 *eis to auto apokathistasas*	497,28 (cf. 499,6.9) *eis tauton apokathistôsas*	*eis tauto apokathistasas* (also F)
1074a14 *men oun*	506,4 *oun*	*men oun*
1074a14 *sphairôn*	506,4 *phorôn*	*sphairôn*
1074a16 [*kai tas aisthêtas*]	506,5 *kai tas aisthêtas*	*kai tas aisthêtas*

Empedocles

DK31	Heiberg	Karsten
B35, line 2 *logou*	529,2 *logôi*	*logou* (F has *logou*)
B35, line 10 *tôn*	529,10 *pô*	*pô* (also F)
B71, line 4 *toss'*	530,4 *toi'* (Stein (1852))	*toss'* (F has *toia*)
B73, line 2 *eidea poipnuousa*	530,7 *aither' epipneiousa* (Stein (1852))	*eidea poipnuousa* (also F)

Euclid, *Elements*

El. vol. 1	Heiberg	Karsten
1, d. 10 (p. 4,1) *poiêi*	538,24 *poiêsêi*	*poiêi* (also F)
3, pr. 19 (p. 216,20) [*gônias*]	538,31 *gônias*	*gônias*
3, pr. 19 (p. 216,21) *eutheia grammê akhthêi*	538,32 *akhthêi eutheia*	*akhthêi eutheia*

Plato, *Phaedo*

Oeuv. Comp. vol. 4.1	Heiberg	Karsten
109A4 *en mesôi*	535,28 *plêsion*	*plêsion*

Plato, *Republic*

Oeuv. Comp. vol. 7.2	Heiberg	Karsten
617A8 *deuterous*	475,16 *deuteron*	*deuterous*

Plato, *Timaeus*

Oeuv. Comp. vol. 10	Heiberg	Karsten
39A3 *periêiein*	475,14 *periietai*	*periêiein* (F has *periêiei*)
40B8 *men*	517,7 omit	*men* (and F)
40C1 *tên*	517,7 omit	omit
40C1 *polon tetamenon*	517,7 *tetamenon polon*	*polon tetamenon*

Timaeus Locris

Marg (1972)	Heiberg	Karsten
215,7 *aôs*	517,24 *augas*	*aous* (also F)
215,8 *anatolas*	517,24 *antolas*	*anatolas* (also F)

(d) Lemmas

Here I bring together places where the text in a lemma printed by Heiberg differs from Moraux's text of Aristotle. I also indicate what is printed by Karsten and (where Heiberg provides the information) what appears in F. In general Heiberg's text reproduces A. I have paid no attention to the numerous differences regarding elision (e.g. *de* vs. *d'*) or minor variations in spelling (e.g. *hauton* vs. *heauton* or *teleiotaton* vs. *teleôtaton*). I should perhaps note that the lemmas in Heiberg and Karsten generally give only the first and last few words of a passage, and so represent less than 10 per cent of the text of *De Caelo*.

Moraux	Heiberg	Karsten
291b17 *d'*	479,1 *de, ei*	*d'* (also D and E; F has *de*)
291b24 *aporiain ousain*	480,24 *aporiôn ousôn*	*aporiain ousain* (also D, E, and F)
292a17 *alogon an doxeien einai*	480,24 *an alogon einai doxeie*	*alogon an doxeien einai* *alogon an doxeien* (D and E) *an alogon doxeien einai* (F)
292a19 *autôn monon*	482,1 *monon autôn*	*autôn monon* (also E) *monon autôn* (F)
292b25 *kinêseôn*	485,4 *tôn kinêseôn*	*kinêseôn* (also D, E, and F)
293a4 *tode*	491,12 *touto*	*tod'* (also D and E; F has *touto*)
293a5 *phorai*	491,12 *sphairai*	*phorai* (also D; F has *sphairai*)
293b30 *apekhontôn*	515,15 *apekhontôn hêmôn*	not in Karsten's lemma (F has *apekhontôn*)
294b23 *tês gês*	525,14 *tautês*	*tês gês*
295a29: *aitia*	525,15 *aition*	*aitia* (also F)
297b17 *te*	542,12 omit	*te*
297b22 *mê ho*	545,29 *mê*	*mê ho* (also F)

English-Greek Glossary

This glossary gives standard Greek equivalents for many nouns, verbs, adjectives, adverbs, and a few prepositions in the translation. Most Greek words which occur only once are omitted, as are many words which have been given a variety of translations (e.g. *logos* and *apodidonai*) and certain very common words (e.g. *einai, ekhein,* and *legein*). The reader will get a better sense of the range of a Greek word by looking at the Greek-English Index for the word and ones closely related to it.

able, be: *dunasthai, endekhesthai*
above: *huperanô*
absurd: *atopos*
accepted, be: *sundokein*
account, take into: *proslogizesthai*
act (v.): *poiein*
action: *praxis*; engage in or perform action: *prattein*; engaging in or involving action: *praktikos*
activity: *energeia*
add: *epagein, prostithenai, suntithenai*
addition: *prosthêkê, prosthesis*
agree: *sunkhôrein*
air: *aêr*
akin: *sungenês*
always: *aei, pantote*
amazing: *thaumastos*
analogous: *analogon*; be analogous: *analogein*
angle: *gônia*
animal: *zôion*
annual: *etêsios*
anomalous: *atopos*
apart: *khôris*
apogee, at: *apogeios*
apparent: *sumphanês*
appear: *phainesthai, phantazesthai*
appearance: *phantasia*
apprehend: *katalambanein*
apprehension: *katalêpsis*
appropriate: *oikeios*
arc: *periphereia*
area: *embadon, khôrion*
argue: *epikheirein, sullogizesthai*
argue (for): *kataskeuazein*
argue against: *antilegein*
argument: *epikheirêma*
argument against: *antilogia*
arrangement: *diathesis*
assimilation: *homoiôsis*
assume: *hupolambanein, lambanein*; assume (first or at the start or previously): *prolambanein*
assumed, be: *hupokeisthai*
assumption: *hupolêpsis*
astronomer: *astrologos, astronomos*
astronomical: *astrologikos, astronomikos*
astronomy: *astrologia, astronomia*
attached, be: *prosgignesthai*
attack (v.): *enkalein*
authoritative: *kurios*
axiom: *axiôma*
axis: *axôn*

backward motion: *hupopodismos*
balanced evenly: *isorrhopos*
base (n.): *basis*
basket-shaped: *kalathoeidês*
belong: *huparkhein*
best: *aristos*
better: *beltiôn, kalliôn, kreittôn*
birth, give: *tiktein*
bisected: *dikhotomos*
block (v.): *antiphrattein, apophrattein*

body: *sôma*
book: *biblion*
bottom: *basis*
bound, be: *dedesthai*
bound (v.): *perainein*
breadth: *diastêma, platos*
brief: *brakhus, suntomos*
bright: *lampros*
bulk: *onkos*; increase in bulk: *onkousthai*

call (v.): *kalein*
call to mind: *mnêmoneuein*
cancel: *aphairein*
carry: *pherein*; carry around: *periagein, peripherein, sumperiagein, sumperipherein*; carry around in the contrary direction: *antiperiagein*
cause: *aitia, aition*
censure (v.): *enkalein*
centre of gravity: *mesos tou barous, mesos kata to baros*
central point: *kentros*
chance (adj.): *tukhôn*
change (n.): *metabolê*
change (trans.): *ameibein, metapherein*; change (intrans.): *metaballein, parallattein*; change place: *metabatikôs kineisthai*; change position: *methistanai*
changing place (adj.): *metabatikos*
choose: *haireisthai*
chosen: *hairetos*
circle: *kuklos*; moving in a circle: *kuklophorêtikos*
circuit: *periodos*
circular: *kuklikos, enkuklios*; circular (of an argument): *diallêlos*
circumference: *periphereia*
clear: *dêlos, enargês, kataphanês, saphês*; make clear: *dêloun*
clearly: *saphôs*
clepsydra: *klepsudra*
clever: *perittos*
close (adj.): *engus*
come to be: *gignesthai*
come together: *sunerkhesthai*
coming to be (n.): *genesis*
common: *koinos*
communicate: *diadidonai*
compare: *paraballein*
comparison: *sunkrisis*
complete (adj.): *pantelês, teleios*; complete (v.): *suntelein*
completion: *apotelesma*
compounded, be: *sunkeisthai*
compress: *sumpilein, thlibein*; compress forcibly: *sunôthein*
concave: *koilos*
conceive: *ennoein, epinoein, hupolambanein, noein*
conception: *ennoia*
concern oneself: *pragmateuesthai, spoudazein*
conclusion: *sumperasma*
condition: *katastasis*
cone: *kônos*
configuration: *skhêmatismos*
confirm: *bebaioûn, pistousthai*
connect: *sunaptein*
consequence, be a: *akolouthein*
consider: *apoblepein, episkeptesthai*
construct: *sunistanai*
contain: *periekhein, perilambanein*
container: *angeion, angos*
containing: *periektikos*
contented, be: *agapân*
contrary motion: *antikinêsis*
converge: *sunneuein*
conversion: *antistrophê*
convert: *antistrephein*
convex: *kurtos*
coordination: *suntaxis*
corporeal: *sômatikos*
correct (v.): *diorthoun*
cosmic order: *diakosmêsis*
counteract: *anelittein*
counterearth: *antikhthôn*
crescent-shaped: *mênoeidês*
cup: *potêrion*
curl: *sustrephesthai*
curved: *peripherês*
cut (v.): *temnein*
cylinder: *kulindros*

dark: *melas*
day: *hêmera*
declare: *apophainesthai*
deficient: *elleiptikos, ellipês*; be deficient: *elleipein*
degree: *moira*
delimit: *diorizein, hôrizein*
demiurge: *dêmiourgos*
demiurgic creativity: *dêmiourgia*
demiurgic: *dêmiourgikos*

demonstrate: *apodeiknunai*
demonstration: *apodeixis*
deny: *apophanai*
depart: *existanai, parakhôrein*
departure: *parakhôrêsis*
depth: *bathos, puthmên*
derivative: *allotrios*
describe: *graphein*
desire (n.): *orexis, hormê*
destroy: *diaballein*
detached: *apêrtêmenos*
determine: *aphorizein, diorizein, hôrizein*
diameter: *diametros*
die (n.): *astragalon*
differ: *diapherein*
difference: *diaphora, parallagê*
different: *diaphoros*
differentia: *diaphora*
difficult: *aporos, ergôdês*
difficulty: *aporia*; raise a difficulty: *aporein*
diopter: *dioptra*
directly: *prosekhôs*
disagree: *diaphônein*
disagreement: *diaphora*
discover: *heuriskein*
discovery: *heuresis*
displaced, be: *parakhôrein*
dissolve: *dialuein*
distance: *apostasis, apostêma, diastasis, diastêma*; be at a distance: *diistanai*
distant, be: *apekhein, aphistanai*
distinguish: *diakrinai, diorizein*
divide: *diairein, diakrinai, diorizein, merizein*
divine: *theios*
division: *diairesis, diakrisis, merismos*
do: *poiein*
dominate: *epikratein, kratein*
domination: *epikrateia*
doubly convex: *amphikurtos*
drum-shaped: *tumpanoeidês*

earlier: *palaios*
earth: *gê*
east (adj.): *anatolikos*; east (n.): *anatolê*
easy: *prokheiros, rhaidios*
eccentric: *ekkentros*
eclipse: *ekleipsis*
be eclipsed: *ekleipein*
elegant: *kharieis*
element: *stoikheion*
elephant: *elephas*
elliptical: *ellipês*
encompass: *perilambanein*
end (n.): *telos*
endure: *diatelein*
entire: *holos*
enunciation: *protasis*
epicycle: *epikuklos*
equal: *isos, homoios*
equalise: *anisazein*
equator: *isêmerinos*
escape notice: *lanthanein*
eternal: *aidios*
even balance: *isorrhopia*
evidence: *tekmêrion*; use as evidence: *tekmairesthai*
evident: *phaneros*
example: *paradeigma*
exceed: *huperairein*
exercise (v.): *gumnazesthai*
exist: *huparkhein, huphistanai*
express: *apangellein*
extend: *ekballein*
extreme, extremity: *akron*
extrude: *exekhein*
extrusion: *exokhê*
eye: *omma, ophthalmos, opsis*

fabricate: *plassein*
fall (v.): *katapheresthai, piptein*
far (away): *porrô*
fast: *takhus*
fastened, be: *endedesthai*
faster: *thattôn*
few: *oligos*
fewer: *elattôn/elassôn*
fiction: *plasma*
figure: *skhêma*
fill: *plêroun*; fill out: *sumplêroun*
find: *exeuriskein, heuriskein*
finish up: *sumperainein*
finite: *peperasmenos*
fire: *pur*
fit in(to): *enarmozein*
fixed: *aplanês*
flat: *platus*; flatness: *platos*
float: *epinêkhesthai*
follow: *akolouthein*
following (adj.): *akolouthos*
foot: *pous*

foot wide (adj.): *podiaios*
force (v.): *biazein*; force (n.): *bia*; forced: *biaios*
form (n.): *eidos*
full moon, at: *panselênos*
furnishing (n.): *euporia*

general: *koinos*
generate: *gennân*
go beyond: *parallattein*
go forward: *proienai*
god: *theos*
good: *agathos*; good itself (n.): *autoagathon*
goodness: *agathotês*
grasp: *ennoein*, *hairein*
great: *megas*
grow: *auxanein*
guard (v.): *phulattein*
guardpost: *phulakê*

hair: *thrix*
hard to move: *duskinêtos*
harmonise: *sumphônein*, *sunarmozein*
hazardous: *parabolos*
health: *hugeia*
hear: *akouein*
heart: *kardia*
heaven: *ouranos*; heavenly: *ouranios*
heavy: *barus*
height: *anastêma*, *hupsos*
hemisphere: *hêmisphairion*
hide from sight: *apokruptein*
high: *meteôros*
hinder: *empodizein*
historical discussion or account: *historia*
history: *historia*
hold: *huparkhein*, *huphistanai*
hold up: *anekhein*
hole: *opê*
hollow (adj.): *koilos*
homocentric: *homokentros*
homoiomerous: *homoiomerês*; homoiomerousness: *homoiomereia*
honourable: *timios*
horizon: *horizôn*
horn-shaped: *keratoeidês*
human (being): *anthrôpos*
hypothesis: *hupothesis*
hypothesise: *hupotithenai*
hypothetical: *hupothetikos*

idea: *epibolê*
illumination: *phôtismos*
illuminated, be: *phôtizesthai*
imagination: *phantasia*
immediate: *amesos*
implausible: *apithanos*
impossible: *adunatos*
inclination: *rhopê*
incline (v.): *rhepein*; be inclined: *enklinesthai*; not inclining in any direction: *aklinês*
increase (v.): *prosauxein*, *auxanein*
indicate: *dêloun*, *endeiknunai*, *epideiknunai*, *sêmainein*
inequality: *anisotês*
inexplicable: *alogos*
infer: *sullogizesthai*
infinite: *apeiros*
inquire: *zêtein*
inquiry: *zêtêsis*
instrument: *organon*
intellect: *nous*
interposition: *epiprosthêsis*
interpret: *exêgeisthai*
interpretation: *exêgêsis*
introduce: *paragein*, *prostithenai*
investigate: *zêtein*
invisible: *adêlos*, *aphanês*
irrational: *alogos*
issue: *problêma*

join: *epizeugnunai*
justification: *pistis*

keep: *phulattein*
kinetic: *kinêtikos*
kinship: *sungeneia*
know: *epistasthai*, *gignôskein*; not know: *agnoein*

lack: *elleipein*
ladle: *kuathos*
large: *megas*
last (adj.): *eskhatos*, *loipos*, *teleutaios*
later: *husteros*, *metagenesteroi*
lazy: *argos*
learn: *manthanein*
leave: *kataleipein*
left behind, be: *hupoleipeisthai*
length: *mêkos*
lentil-shaped: *phakoeidês*
less: *elattôn*/*elassôn*, *hêtton*

lie above: *huperkeisthai*
life: *zôê*; lifeless, without life: *azôs*
light (adj.): *kouphos*
light (n.): *phôs*
likely: *eikos*
limit (n.): *peras*
limit (v.): *apostenoun*
line: *grammê*
listen to: *peithesthai*
little: *brakhus, mikros*
live: *oikein*
Love (n.): *Philia, Philotês*
lunar: *selêniakos*

magnitude: *megethos*
maintain: *phulattein*
make: *parekhein, poiein*
manifest: *prodêlos*
manner: *tropos*
many times as great or large: *pollaplasios*
mass, having: *athroos*
mathematician: *mathêmatikos*
mathematics: *mathêmata*
matter: *hulê*
mean (v.): *dêloun, sêmainein*
means: *hodos*
measure, measurement (n.): *metron*
measure (v.): *metrein*
mention: *hupomimnêskein, mnêmoneuein*
meridian: *mêsembria*
milk: *gala*
mix: *mignunai*
mode: *tropos*
monad: *monas*
month: *meis*
moon: *selênê*
motion: *kinêsis, phora*; motionless: *akinêtos*
mountain: *oros*
mouth: *stomion*
move (intrans.): *kineisthai, pheresthai, khôrein, metabainein*; move (trans.): *kinein*; move along with: *sunkineisthai*; move in the contrary direction: *antipheresthai*; move in the same direction: *sumpheresthai*; move out: *exerkhesthai*; move under or down: *huperkhesthai*; move under: *hupotrekhein*
multitude: *plêthos*
myth: *muthos*
mythical: *muthikos*

natural: *phusikos, kata phusin*
nature: *phusis*
near: *plêsios, prosêkhês*; be or get near: *plêsiazein*
necessary: *anankaios*
necessity: *anankê*
next: *ephexês, loipos*
non-uniform: *anômalos*; non-uniformity: *anômalia*
north (adj.): *arktos, boreios*
note (v.): *ephistanai*
number: *arithmos, plêthos*

object (v.): *enistanai*
objection: *enstasis*
oblique: *loxos*
obscure (v.): *epiprosthein*
observe: *paraphulattein, têrein, theasthai*
obvious(ly): *dêladê, dêlonoti, epidêlos, prokheiros*
one's own: *oikeios*
opposite: *antistrophê, enantios*; be opposite: *antikeisthai*
opposition: *enantiôsis*
order (n.): *suntaxis, taxis*
order (v.): *diatassein*
ordinary: *koinos*
organ: *organon*
ought: *opheilein*

parallel: *parallêlos*
parallelogram: *parallêlogrammon*
part: *meris, meros, morion*
partial, particular (adj.): *merikos*
partless: *amerês*
pass through: *khôrein*
passage, passageway: *parodos*
path: *hodos*
peculiar: *idios*
penetrate: *diikneisthai*
perceptible: *aisthêtos*
perception: *aisthêsis*
perfect (adj.): *pantelês*
perigee, at: *perigeios*
perimeter: *perimetros*
perpendicular: *kathetos, orthos*
phase: *phasis*
philosophical: *philosophos*; subject of

philosophical discussion: *philosophêma*
pillar: *kiôn*
place (v.): *tithenai*
place (n.): *topos*
plane (figure): *epipedos*
planet: *planetês*
plausible: *pithanos*
poet: *poiêtês*
point (n.): *sêmeion, stigmê*
pole: *polos*
portion: *moira*
posit (v.): *tithenai*
position: *topos*
possible: *dunatos*; be possible: *dunasthai, endekhesthai*
posterior: *husteros*
power: *dunamis*
practically: *skhedon*
precede: *proêgeisthai*; preceding: *prosekhôs*
precise: *akribês*
predicate (v.): *katêgorein*
predominance: *epikrateia*
premiss: *lêmma, protasis, protethen*
preserve: *aposôzein, diasôzein, sôzein*
prevent: *kôluein*
primary, be: *proêgeisthai*
principle: *arkhê*
prior, be: *huperairein*
problem: *problêma*
proceed: *proerkhesthai*
procession: *proödos*
produce: *apotelein, gennân, poiein*
product: *apotelesma*
progress: *proïenai*
progression: *akolouthêsis*
project (v.): *proballesthai*; project upward: *anarrhiptein*
prominent, be: *epikratein*
proof: *deixis*
proper: *idios, oikeios*
proportion, proportionality: *analogia*; proportional: *analogon*
propose: *epiballein, proballesthai, protithesthai*
prove: *deiknunai, paradeiknunai*
provide: *parekhein*
proximate: *prosêkhês*
proximity: *geitniasis*
purpose: *skopos*
push: *ôthein*
push out: *exôthein*
put forward: *proballesthai*

quick: *takhus*

raise up: *meteôrizein*
random: *tukhôn*
rational: *logikos*
ray: *aktis*
reach: *aphikneisthai, hêkein*
reason: *aitia*; reasonable: *eikos, eulogos, kata logon*
recognise: *ephistanai, ephistanein, gignôskein*
record, recount: *historein*
refer: *mnêmoneuein*
reflect: *sunnoein*
refutation: *antilogia*
refute: *anatrepein, dielenkhein, elenkhein*
region: *khôra, topos*
reject: *diaballein*
relation: *skhesis*
release: *aphienai, luein*
remain: *diatelein*; remain (fixed or at rest): *menein*; remaining: *loipos*
remind: *hupomimnêskein*
resemble: *eiokenai*
resist: *antikoptein, antereidein*
responsible: *aitios*
rest (n.): *êremia, monê*
rest (v.): *êremein*
restoration: *apokatastasis*; simultaneous restoration: *sunapokatastasis*
restored, be: *apokathistasthai*; be restored simultaneously: *sunapokathistasthai*
retain: *phulattein*
retardation: *hupoleipsis*
retrogression: *proêgêsis*
revolution: *periagôgê, periphora, strophê*; make a revolution: *periienai*
right: *dikaios, orthos*
rise: *anatellein*; rise up: *epipolazein*; rising (n.): *anatolê*
room: *khôra, topos*
rotation: *peristrophê, sustrophê*
round, roundish: *kukloterês, peripherês*

same: *homoios*; much the same: *paraplêsios*

scribal error: *graphikon ptaisma*
search (v.): *zêtein*
section: *apotomê*
see: *apidein, horân, theôrein*
seek: *zêtein*
seem: *eoikenai*
separate (v.): *khôrizein*
serviceable: *khrêsimos*
set (v.): *tithenai*
set out: *ektithesthai, paratithesthai, protithenai*
setting (n.) *dusis*
shadow: *skia*; covered in shadow: *skieros*
shape: *skhêma*
share in: *metalambanein*
show (v.): *dêloun, endeiknunai, epideiknunai*
side (adj.): *plagios*
sight: *opsis*
similar: *homoios, paraplêsios*
simple: *haplos*
sit like a lid: *epipômatizein, epipômazein*
situated, be: *hidrusthai*
size: *megethos*
slight: *brakhus, oligos*
slow: *argos, bradus*
small: *brakhus, mikros*
smaller: *elattôn/elassôn*
smooth: *katallêlos*
solar: *hêliakos*
solid (adj.): *stereos*
solution: *lusis*
solve: *dialuein, luein*
soul: *psukhê*; having or involving soul (adj.): *empsukhos*; without soul: *apsukhos*
south (n.): *mesêmbria*
south, southward: *notios*
sovereign: *arkhikos*
space: *khôra*
speak against: *anteipein*
special character: *idiotês*
speed: *takhos*; having the same speed (adj.): *isotakhês*
sphere: *sphaira*; description of the sphere: *sphairopoiia*; production of a sphere: *sphairôsis*; spherical: *sphairikos, sphairoeidês*
spill over: *huperekkheisthai*
split apart: *diistanai*
stade: *stadion*
stand (still): *histanai*
stand in font of: *epiprosthein*
standing still (n.): *stêrigmos*
star: *astêr, astron*; starless: *anastros*
start (n.): *arkhê*; start (v.): *arkhein, hormân*; starting point: *aphormê, arkhê*
station: *stêrigmos*
stir up: *hormân*
straight: *euthus*
strange: *atopos*
stretch (v.): *epiteinai*; stretching (n.): *tasis*
Strife: *Neikos*
strive: *speudein, spoudazein*; striving (n.): *spoudê*
strong: *enkratês, iskhuros, karteros*; stronger: *kreittôn*
structure: *sustasis*
study (n.): *theôria*
study (v.): *theôrein*
substance: *ousia*
subtract: *aphairein*
successful, be: *katorthoun*
sufficient: *autarkês, hikanos*; be sufficient: *arkein*
suitable: *emmelês, epitêdeios*; be suitable: *prosêkein*
superiority: *huperokhê*
support (v.): *hupereidein, okhein*
surface: *epiphaneia*
surprising: *thaumastos*
surprised, be: *thaumazein*
surround: *periekhein*
surrounding: *perix*
swell (v.): *kumainein*
syllogism: *sullogismos*; produce a syllogism: *sullogizesthai*
system: *suntaxis*

tangent, be: *ephaptesthai*
tear apart: *diaspân*
text: *graphê, lexis*
theorem: *theôrêma*
theory: *theôria*
thesis: *thesis*
three-dimensional: *stereos*
time: *khronos*; taking the same amount of time (adj.): *isokhronios*
traversal: *diexodos*
traverse: *diexienai, diienai*
treatise: *sungramma*
true: *alêthês*

truth: *alêtheia*
try: *enkheirein, epikheirein, peiran*
turn (intrans.): *strephesthai, poleuein, metienai, metabainein*; turn (trans.): *epistrephein*; turn in the same direction (trans.): *sunepistrephein*
turning away (n.): *ektropê*
twinkle: *stilbein*

ultimate: *eskhatos*
unclear: *adêlos, asaphês*
uncountable: *anarithmêtos*
undergo: *paskhein*
underlie: *hupokeisthai*
understand: *akouein, ekdekhesthai, eklambanein*
unification: *henôsis*
uniform: *homalês*; uniformity: *homoiotês*
unify: *henoun*
unique: *idios*; uniqueness: *idiotês*
unite: *henoun*
unity: *henôsis*
universal: *holikos, katholikos, katholou*
unmixed: *amiktos, eilikrinês*
unmoving: *akinêtos*
unnatural: *para phusin*
unreasonable: *alogos*
unsolvable: *alutos*
upright: *orthos*
urgent, be: *katepeigein*
use (n.): *khreia*
use (v.): *khrêsthai, proskhrêsthai*

validate: *epideiknunai*
variegated: *poikilos*
vertex: *koruphê*
vision: *opsis*
vitally: *zôtikôs*
void (adj.): *kenos*
voluntary: *hekôn, hekousios*
vortex, vortex motion: *dinê, dinêsis*

wander: *planasthai*
wane (v.): *phthinein*
water: *hudôr*
wax: *auxanein*
way: *tropos*
way out: *diexodos*
weight: *baros*; having weight (adj.): *barus*
west (n.): *dusis, dusmê*
whole (adj.): *holos*
wind (n.): *anemos*
wind, be wound: *illesthai*
winter (adj.): *kheimerinos*
witness (n.): *marturia, marturion*
wonderful: *thaumastos*
word: *epos, lexis, rhêma*; words: *lexis, rhêsis*
worse: *katadeesteros, kheirôn*
worth (adj.): *axios*
worth (n.): *axiôma*
write: *graphein*

year: *etos*

zenith: *koruphê*
zodiac: *zôidiakos kuklos, to zôidiakon*; sign of the zodiac: *zôidion*

Greek-English Index

This index, which is based on Heiberg's text with my emendations, gives the English translations of many nouns, verbs, adjectives, adverbs, and a few prepositions used by Simplicius; certain very common words (e.g. *einai*, *ekhein*, and *legein*) and number words are omitted, as are words which only occur in quotations (or apparent quotations) of other authors (the exceptions to this are Alexander, Ptolemy, and Sosigenes). When a word occurs no more than ten times, its occurrences are listed; in most other cases only the number of occurrences is given. Occurrences in lemmas are ignored. Sometimes comparatives, superlatives, and adverbs are included under the positive form of an adjective, sometimes they are treated separately. There is a separate index of names.

Index of Passages

(a) Testimonia and fragments

I list here passages from Simplicius which have been collected as testimonia about or fragments of various ancient authors.

(b) Texts quoted, closely paraphrased, or clearly referred to by Simplicius or Alexander

(c) Early texts cited in the notes

References are to the line in the Greek text on which a footnote number occurs.

Index of Names

In many cases information on an item or reference to where information can be found is provided in the note on a given passage. For sun and moon see the English-Greek Glossary and the Greek-English Index; see also the Introduction. For Alexander in the index of individuals and groups (section c) I indicate, where possible, Simplicius' assessment of what Alexander says; for Aristotle I list passages in which something general is said about him or there is a reference to material outside *On the Heavens*. The index of modern scholars (section d) does not include editors of texts unless they are mentioned for their position on an editorial or interpretive issue; reference to a page and line indicate the position of a note in which the scholar in question is mentioned.

(a) Astronomical names

(b) Geographical names

(c) Individuals and groups

(d) Modern scholars cited in the notes

Subject Index

This index lists places where Simplicius' discussion goes beyond straightforward exposition of Aristotle's text. See also the other indices and the English-Greek Glossary.

www.ingramcontent.com/pod-product-compliance
Ingram Content Group UK Ltd.
Pitfield, Milton Keynes, MK11 3LW, UK
UKHW021239300926
15471UKWH00013B/676